Introduction

The aim of this book is to make diagnosing and fault finding easier. It has been designed to help you by giving you all the information you need in a clear and logical manner.

We have worked closely with many manufacturers training and service departments to bring their knowledge to you.

 https://www.youtube.com/user/mrcombitraining/videos

 https://www.facebook.com/mrcombitraining/

https://twitter.com/MrCombiTraining

Please use this book as a guide, ALWAYS get advice before you start from the manufacture and have the Installation & Service manual.

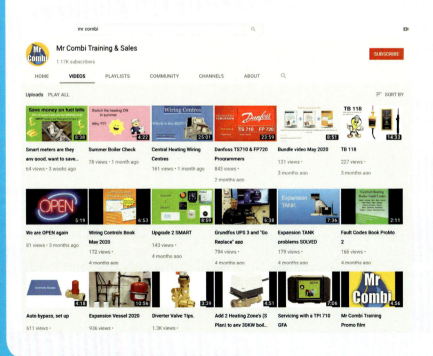

Watch videos on our channel for more information.

When you see the YouTube logo it means there is a video to support it.

Table of Contents

"Read this first"

All testing MUST be made by a suitably qualified Gas Safe engineer or electrician

Pump: **NEVER** remove the screw to spin the impeller to free when stuck, it's a **venting** screw.

Fan / APS Switch: **NEVER** blow to test it, the fan and a digital manometer does that.

Neon screwdriver: **DO NOT** use to test voltage, **YOU** could become the earth that's **DANGEROUS** use a Cat.3 multimeter set to AC see page 16 Polarity Test 1.

Most components require 2 elements, electricity, either gas-water or air, if it has both and fails to respond replace, e.g. a pump has voltage and water do NOT put a screwdriver in the vent screw and spin it manually and now "fixed it" find the real reason, sludge high winding resistance (over 400Ω) or over 5 years old.

Noisy fan: see page 27 & 66

Cold rads: see page 28-29

Noisy boiler: see pages 27, 30, 38 & 39

Pressure loss: see page 27 & 72-76

Gas valve test: see pages 31

Boiler not responding: see pages 32-35

No heating or hot water: see pages 32 & 68-70

Poor hot water flow or temperature: see page 43, 57, 62 & 64

If you smell gas, don't switch anything Off or On tell others, open windows exit the building and

call **0800 111 999**

Don't Panic!

Every successful diagnostic engineer knows all boilers / systems has a sequence, slow down, take your time, don't guess a fault

Before you start do the basic tests first.

1. Is the 3 amp fuse OK
2. Is the system full of water. (1 bar)
3. Is the gas ON (20mBar) standing
4 Is the flue clear

Winter period

5. Is the condense pipe dry, check for frozen water (test by pouring a little water via the analyser test point see M.I.)

CW01335092

Testing in Heating mode first

see page Sequence of Operation

Most boilers before firing up may do a self - diagnostic program, which can repeat every 24 hours. The fan may run at slow speed to check the main heat exchanger is air tight and flue stat / sensor responds to the change of temperature. The diverter may move back and forth.

Next stage is the pump will start and within 30 mili seconds must reach full speed to test the water pressure switch / sensor, if OK the fan will start the same test.

Finally if the high limit, boiler stat and room stat are OK the gas with ignition box will start and the boiler fires up. Within 30 seconds the PCB will re-test all the sensors for a change in value in Ohms and Hz if OK heating will continue as normal, NB some may work on low gas for 3 minutes.

!!! Some combies may have a water filter.

Some spares stockiest may have old stock, always check for upgrades with the factory spares department.

Avoid general Forums lots of misinformation only use Factory Forums. **Find the local boiler expert from the factory web site, they are the best and you can trust them.** Hire them it's quick and you learn

HEATING ON WHEN NOT REQUIRED
(programmer indicates off)

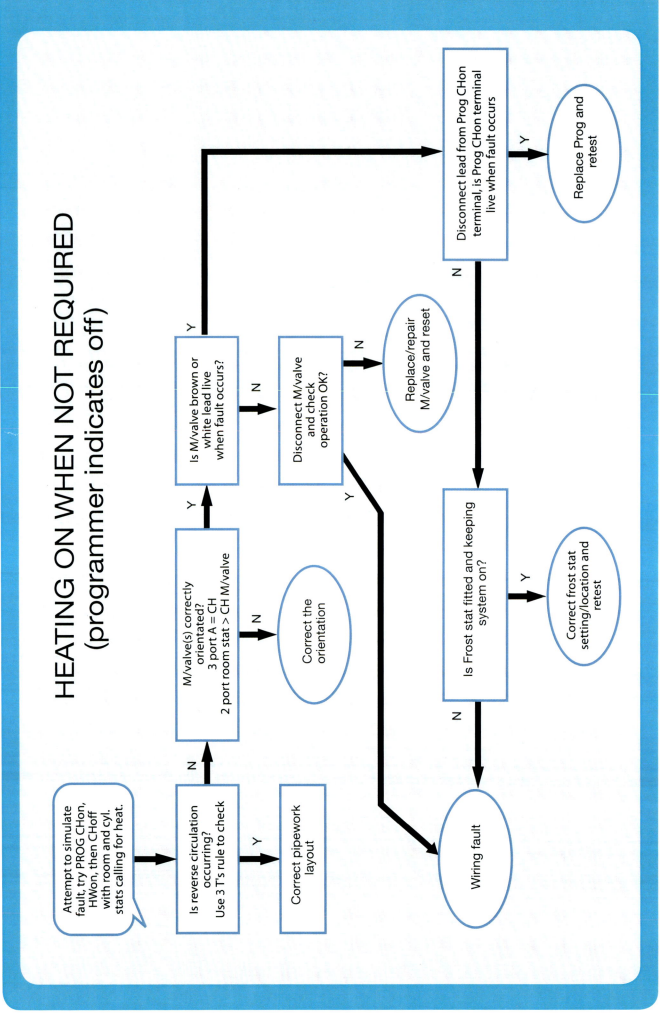

Attempt to simulate fault, try PROG CHon, HWon, then CHoff with room and cyl. stats calling for heat.

Is reverse circulation occurring? Use 3 T's rule to check

Correct pipework layout

M/valve(s) correctly orientated? 3 port A = CH 2 port room stat > CH M/valve

Correct the orientation

Is M/valve brown or white lead live when fault occurs?

Disconnect M/valve and check operation OK?

Replace/repair M/valve and reset

Disconnect lead from Prog CHon terminal, is Prog CHon terminal live when fault occurs

Replace Prog and retest

Is Frost stat fitted and keeping system on?

Correct frost stat setting/location and retest

Wiring fault

NO HOT WATER AVAILABLE

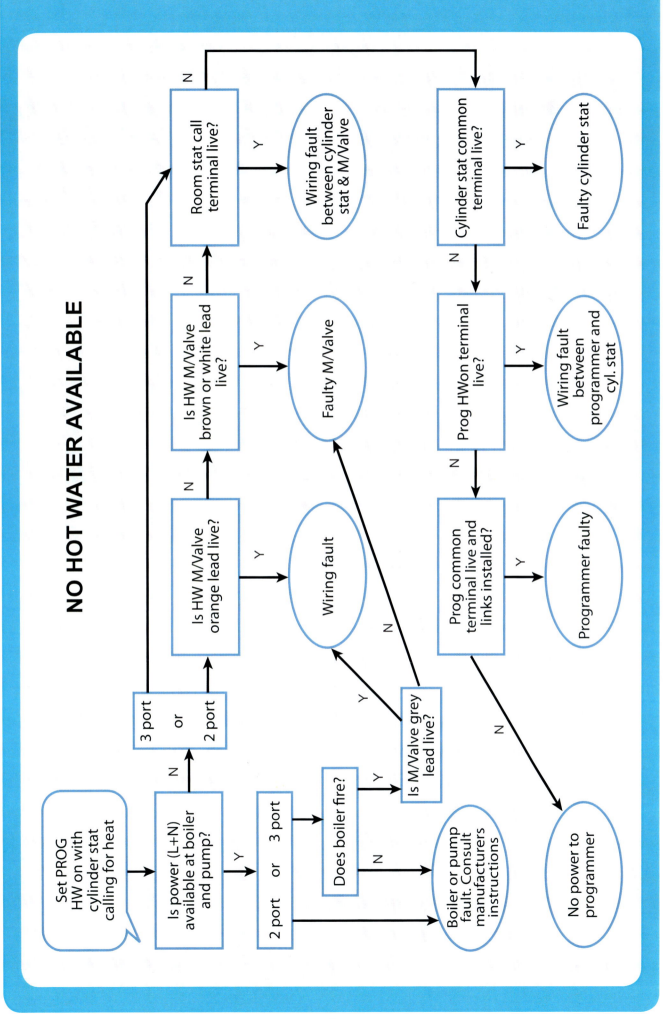

Set PROG HW on with cylinder stat calling for heat

Is power (L+N) available at boiler and pump? — N → 3 port or 2 port

Is power (L+N) available at boiler and pump? — Y → 2 port or 3 port

2 port or 3 port → Does boiler fire?

Does boiler fire? — Y → Is M/Valve grey lead live?

Does boiler fire? — N → Boiler or pump fault. Consult manufacturers instructions

2 port → Boiler or pump fault. Consult manufacturers instructions

Is M/Valve grey lead live? — Y → Wiring fault

Is M/Valve grey lead live? — N → Faulty M/Valve

Is HW M/Valve orange lead live? — Y → Wiring fault

Is HW M/Valve orange lead live? — N → Is HW M/Valve brown or white lead live?

Is HW M/Valve brown or white lead live? — Y → Faulty M/Valve

Is HW M/Valve brown or white lead live? — N → Room stat call terminal live?

Room stat call terminal live? — Y → Wiring fault between cylinder stat & M/Valve

Room stat call terminal live? — N → Cylinder stat common terminal live?

Cylinder stat common terminal live? — Y → Faulty cylinder stat

Cylinder stat common terminal live? — N → Prog HW on terminal live?

Prog HW on terminal live? — Y → Wiring fault between programmer and cyl. stat

Prog HW on terminal live? — N → Prog common terminal live and links installed?

Prog common terminal live and links installed? — Y → Programmer faulty

Prog common terminal live and links installed? — N → No power to programmer

HEATING TO HOT

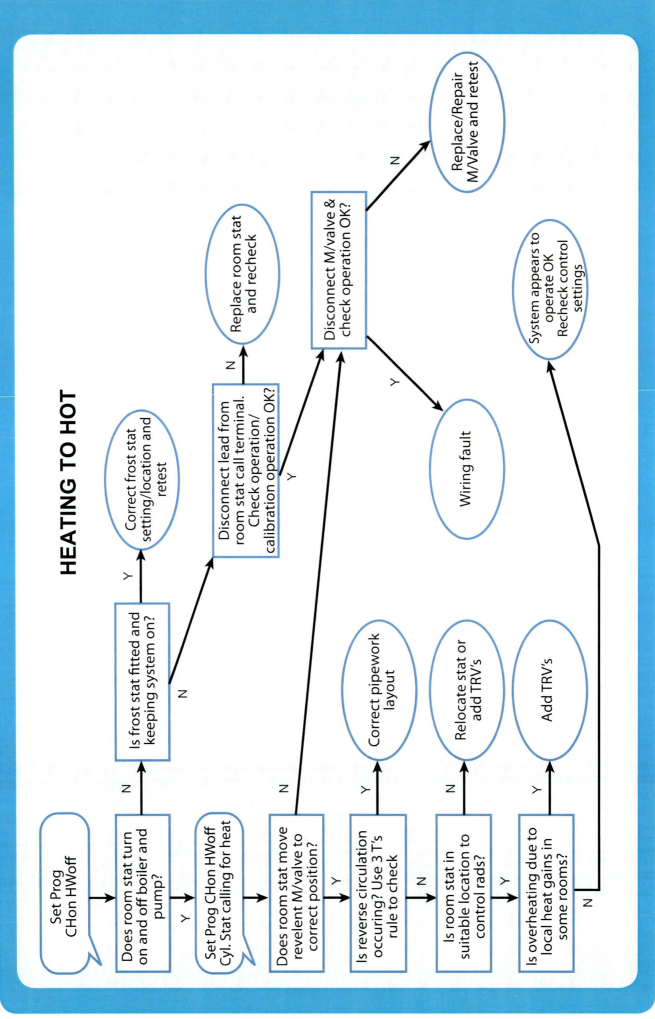

Set Prog CHon HWoff

Does room stat turn on and off boiler and pump?

N → Is frost stat fitted and keeping system on?

Y → Correct frost stat setting/location and retest

N → Disconnect lead from room stat call terminal. Check operation/calibration operation OK?

N → Replace room stat and recheck

Y → Disconnect M/valve & check operation OK?

N → Replace/Repair M/Valve and retest

Y → Wiring fault

Set Prog CHon HWoff Cyl. Stat calling for heat

Does room stat move revelent M/valve to correct position?

N → Disconnect M/valve & check operation OK?

Y → Is reverse circulation occuring? Use 3 T's rule to check

Y → Correct pipework layout

N → Is room stat in suitable location to control rads?

N → Relocate stat or add TRV's

Y → Is overheating due to local heat gains in some rooms?

Y → Add TRV's

N → System appears to operate OK Recheck control settings

NO HEATING AVAILABLE

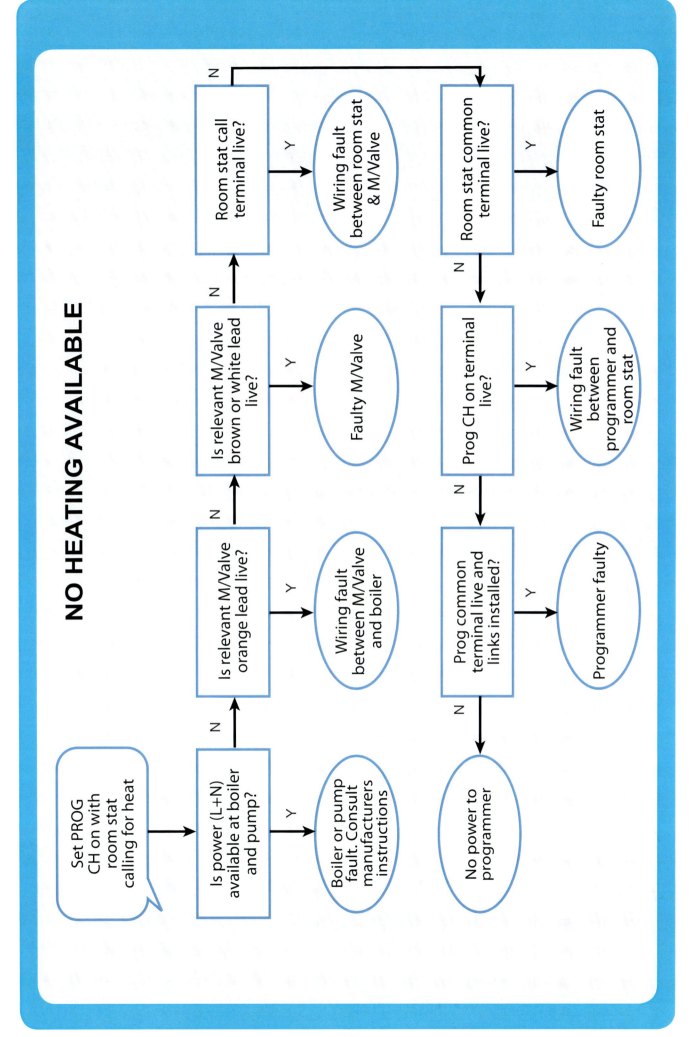

Set PROG CH on with room stat calling for heat

→ Is power (L+N) available at boiler and pump?

— Y → Boiler or pump fault. Consult manufacturers instructions

— N ↑

Is relevant M/Valve orange lead live?

— Y → Wiring fault between M/Valve and boiler

— N ↑

Is relevant M/Valve brown or white lead live?

— Y → Faulty M/Valve

— N ↑

Room stat call terminal live?

— Y → Wiring fault between room stat & M/Valve

— N →

Room stat common terminal live?

— Y → Faulty room stat

— N ↓

Prog CH on terminal live?

— Y → Wiring fault between programmer and room stat

— N ↓

Prog common terminal live and links installed?

— Y → Programmer faulty

— N ↓

No power to programmer

The 4 Steps to Electrical Safety Testing

1. Earth continuity check

2. Short circuit check

3. Resistance to earth check

4. Polarity check

It's very important to check the electrics, before you switch ON which might blow a fuse and cause more damage. Extra test, in Ohms put the black lead on the PCB Earth and the red lead to Live then Neutral terminals both MUST read O.L any numbers or 000.0 means danger there is a big short circuit, see page 14

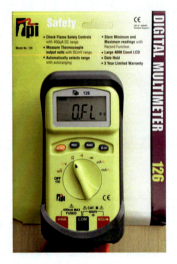

While most people are aware of the danger from electric shock, few realise how little current and how low a voltage are required for a fatal shock.

At about 10 mA, muscular paralysis of the arms occurs cannot release grip.

At about 30 mA, respiratory paralysis occurs, breathing stops and the results are often fatal.

At about 75 to 250 mA, for exposure exceeding five seconds, ventricular fibrillation occurs, causing incoordination of the heart muscles; the heart can no longer function. Higher currents cause fibrillation at less than five seconds. The results are fatal. See page 16-17 Polarity

Only use a CAT. 111 multimeter

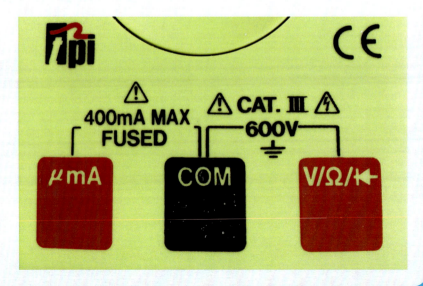

Earth Continuity Test

The earth continuity test ensures the boiler has a good connection to the mains earth supply. This ensures that if a fault arises the boiler is still safe to touch.

If the boiler is not correctly earthed it could cause an electric shock.

1. The boiler should remain switched OFF at the spur/unplugged.
2. Select the Resistance (Ω) setting on the multimeter.
3. Place the black lead onto the casing of the boiler.

EARTH CONTINUITY

Mr Combi Training

4. For boilers with a plug, place the red lead onto the earth pin of the plug:

EARTH CONTINUITY

Mr Combi Training

5. For boilers wired to a spur, place the red lead onto the spur retaining screw:

6. The resistance must be 0Ω If it is higher than 1Ω a fault is present, this must be rectified immediately

Short Circuit Test

This test can determine whether a short circuit exists on the boiler. A short circuit may cause blown fuses and burn damage to the appliance.

Ensure a 3 amp fuse is fitted to all Combi boilers to reduce the effects of a short circuit fault to a minimum.

1. The boiler should remain switched OFF at the spur/unplugged.
2. Switch ON the boiler from the control panel (if a switch is fitted) and turn on all controls so a demand is present.
3. Select the Resistance (Ω) setting on the multimeter.

4. For boilers with a plug place the black lead onto the neutral pin on the plug and the red lead onto the live pin:

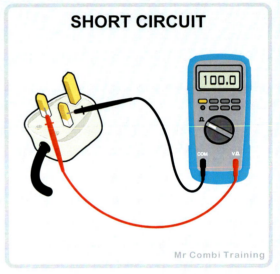

5. For boilers wired to a spur the test must be carried out at the terminal block. Place the black lead into the neutral (blue) terminal and the red lead into the live (brown) terminal:

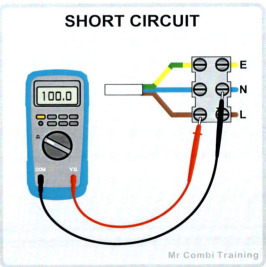

6. The resistance should be at least 100Ω. A lower resistance may indicate a fault, please contact the manufacturer for guidance. 0Ω indicates a short circuit.

To help identify the cause of a fault, circuits within the appliance should be isolated one by one and the test repeated until it clears. See the fault isolation page for guidance.

Resistance to Earth Test

The resistance to earth test ensures the boiler is insulated between live and earth.

If the boiler is not insulated it could cause an electric shock.

1. The boiler should remain switched OFF at the spur/unplugged and ON at the control panel with a demand present.
2. Select the Resistance (Ω) setting on the multimeter.
3. Place the black lead onto the casing of the boiler.
4. Place the red lead into the live (brown) terminal.

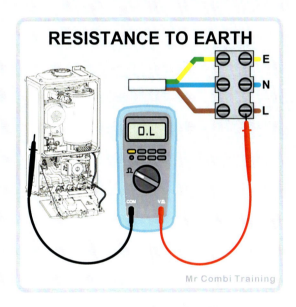

5. The resistance should be at least 1MΩ, Open Loop (O.L.) or display a 1 (depending on your multimeter). A lower resistance indicates a fault. 0Ω indicates a short circuit.

To identify the cause of a fault, circuits within the appliance should be isolated one by one and the test repeated until it clears. See the fault isolation page for guidance.

Polarity Test

This test determines whether the supply electrics have been wired correctly and leakage voltage on the neutral is within limits. If the live and neutral term nals are swapped or the leakage voltage is too high, the boiler may produce unexpected results or fail to switch on.

There are three parts to this test:

1. Live to Neutral > 230vAC
2. Live to Earth > 230vAC
3. Neutral to Earth 00.0vAC

This is a LIVE test, with power applied, so care should be taken.

1. Turn OFF the boiler and controls at the boiler control panel.
2. Switch ON the spur/plug the boiler in.
3. Select the Volts AC (~) setting on the multimeter.

Part 1. Place the black lead into the neutral (blue) terminal. Place the red lead into the live (brown) terminal. The voltage should be the supply voltage of 230 240V.

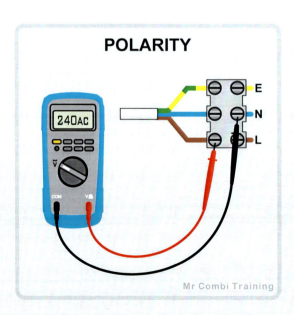

Part 2. Place the black lead onto the casing of the boiler. Place the red lead into the live (brown) terminal. The voltage should be the supply voltage of 230-240V.

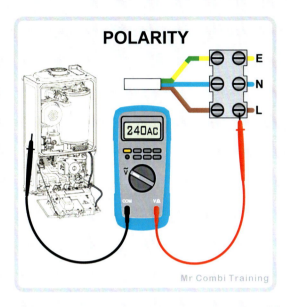

Part 3. Keep the black lead on the bare metal of the casing or where the earth wires meet in the boiler. Place the red lead into the PCB neutral (blue) terminal. The voltage MUST be 0.00v AC if higher a Part P qualified electrician must be called. If the voltage is between 5v AC and 15v AC the boiler may still operate intermittently or not at all.

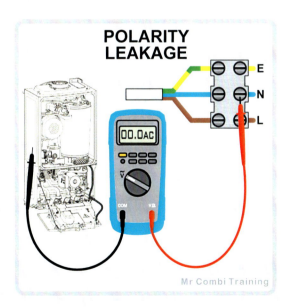

Isolating the Appliance

Before carrying out the electrical safety tests it's important to ensure the boiler is isolated from the power supply correctly.

An incorrectly wired spur could cause the appliance to be powered even when the switch is off.

If the boiler is fitted with a plug it is very simple to isolate by removing the plug from the socket. The plug should then be examined internally to ensure it is wired correctly and a 3 amp fuse is fitted. The simplest way to isolate a boiler with a spur is by turning off the switch and then removing the fuse. Again, the fuse should be rated at 3 amps.

External controls to the boiler may be independently powered or a fault may exist in the spur, this means that the boiler terminal block must be checked for power.

1. Remove the boiler front panel and gain access to the terminal block.
2. Select the Volt AC (V~) setting on the multimeter.
3. Place the black lead into the neutral (blue) terminal.
4. Place the red lead into the live (brown) terminal.

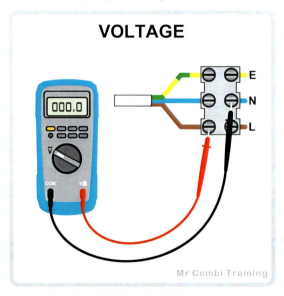

5. The voltage should be 0V. If any voltage is present a fault may exist in the spur or power is available from elsewhere.

A voltage pen can also be used to carry out this test at the terminal block by simply holding it close to the live and neutral terminals in turn.

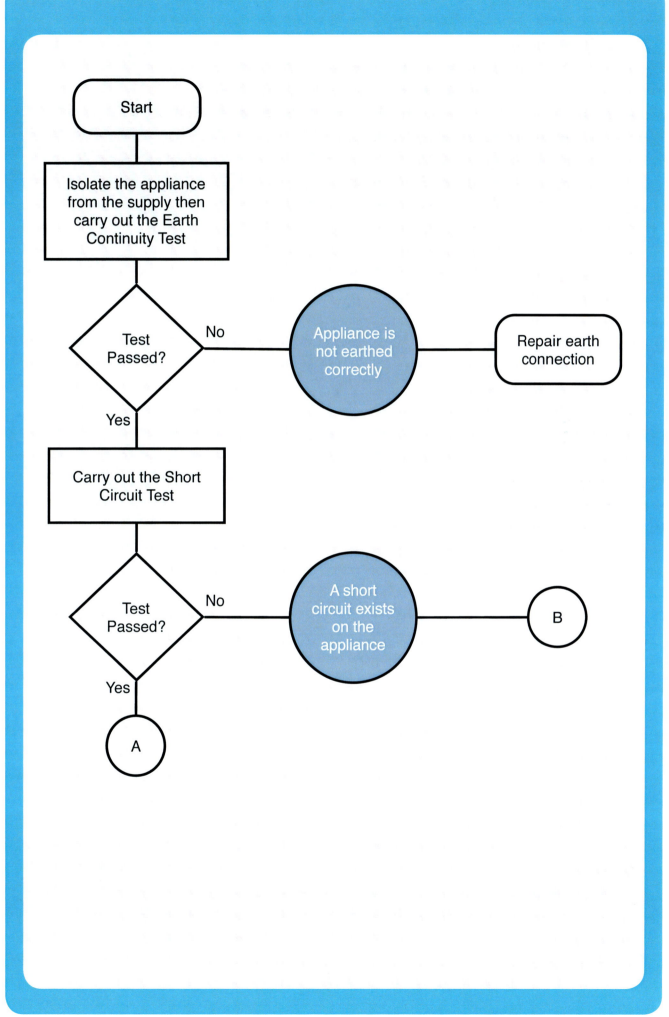

Start

Isolate the appliance from the supply then carry out the Earth Continuity Test

Test Passed?

No → Appliance is not earthed correctly → Repair earth connection

Yes

Carry out the Short Circuit Test

Test Passed?

No → A short circuit exists on the appliance → B

Yes

A

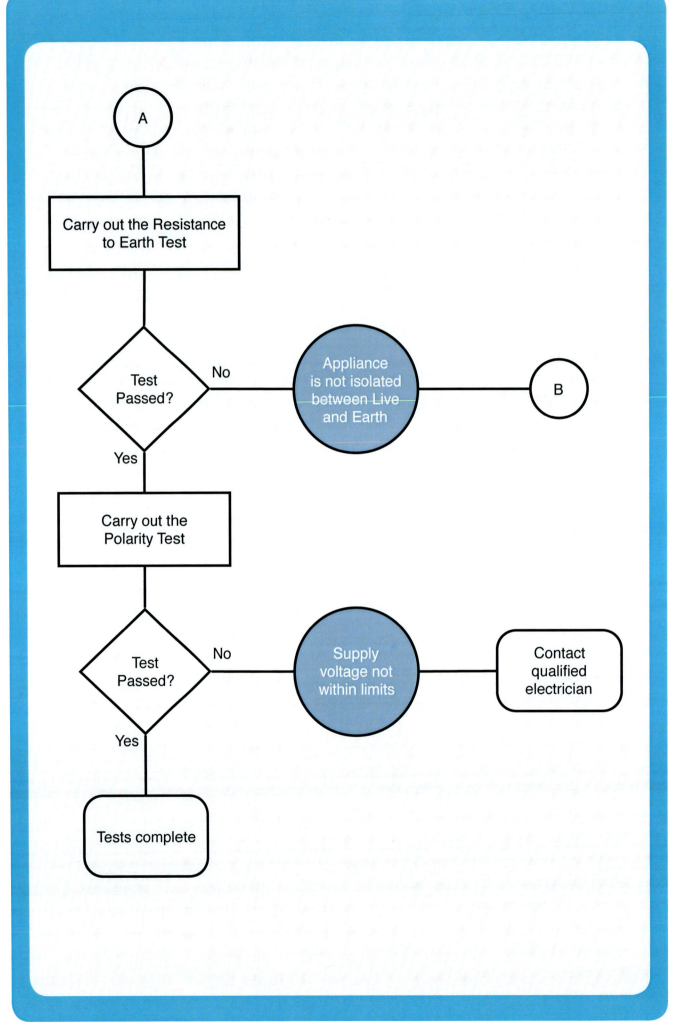

A

Carry out the Resistance to Earth Test

Test Passed?

No → Appliance is not isolated between Live and Earth → B

Yes

Carry out the Polarity Test

Test Passed?

No → Supply voltage not within limits → Contact qualified electrician

Yes

Tests complete

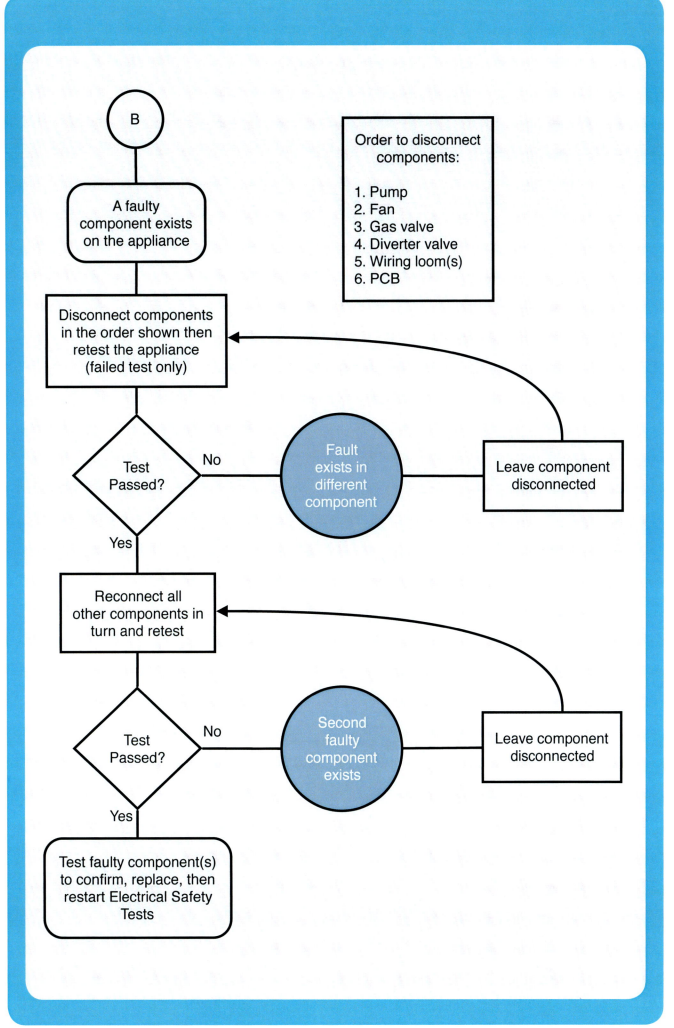

B

A faulty component exists on the appliance

Disconnect components in the order shown then retest the appliance (failed test only)

Order to disconnect components:

1. Pump
2. Fan
3. Gas valve
4. Diverter valve
5. Wiring loom(s)
6. PCB

Test Passed?

No → Fault exists in different component → Leave component disconnected

Yes

Reconnect all other components in turn and retest

Test Passed?

No → Second faulty component exists → Leave component disconnected

Yes

Test faulty component(s) to confirm, replace, then restart Electrical Safety Tests

Fault Isolation

This test determines whether the supply electrics have been wired correctly and leakage voltage on the neutral is within limits. If the live and neutral terminals are swapped or the leakage voltage is too high, the boiler may produce unexpected results or fail to switch on.

If an appliance fails either the Short Circuit Test or Resistance to Earth Test the following sequence can be used to help identify the cause of the fault. The flow chart at the start of this section may help you understand this process better.

To isolate the fault you must disconnect the major components of the boiler one by one until the fault clears. Start by disconnecting the pump from the loom as this is the most common item to fail. The test should then be repeated.

Test Pass:

The pump is likely to be at fault and should be tested for confirmation.

Test Fail:

Disconnect the next component (without reconnecting the pump) and repeat the process.

The boiler major components should be isolated in the following order:

1. Pump

2. Fan

3. Gas Valve

4. Diverter Valve

5. Wiring Loom(s)

6. PCB (see the next section for further guidance)

Once the faulty component has been isolated, reconnect all other components one-by-one and retest to ensure only one fault exists. After replacing a part, ensure the Electrical Safety Tests are carried out fully before turning on the appliance.

Hints and Tips

Below are some general hints and tips that you should follow to keep yourself safe and protect any appliances from damage.

Before you begin work

- Always do a risk assessment before starting any job.
- Check the boiler is fitted with a 3 amp fuse.
- We recommend fitting boilers with a fused plug, spurs do go faulty and are harder to isolate.
- Use a good auto-ranging multimeter and volt pen.
- Check your tools regularly to ensure that they're fit for purpose.
- Carry out the four part Electrical Safety Tests before working on any appliance.
- Before opening up an appliance, ensure it's safely isolated and check there is 0 Vac between live and neutral on the PCB.
- Ensure the spur or un-switched socket for the boiler.

Common faults on boilers and circuits

- Under sized gas pipes - There is no 15mm gas supply for condensing boilers, only 22, 28 or 34mm.
- A 13 amp fuse has been fitted instead of 3 amp.
- No neutral on a 230V room stat.
- No frost stat fitted to a boiler installed in the loft / garage etc. - The boiler may have one but the system also needs one to protect the pipes.
- No pipe stat fitted where a frost stat has been used -This is a common fault.
- No or incorrect by-pass on the system - The boiler may have one the system also needs one.
- Expansion vessel is too small or flat.
- Poor pipe / tank insulation.
- Incorrect gas supply pressure - Must be 18 mbar working pressure at the gas valve. If not call the manufacture for guidance.
- No or incorrect air vents.
- Improper flue termination.
- Unsupported flex or pipes.
- Incorrect cable to the boiler - **Must be 0.75 mm flex.**
- Loft tanks not checked on annual service.

Sequence of Wiring

One of the main reasons why there are so many errors in wiring up systems is because installers don't have a set sequence. We have therefore developed a sequence which is easy to follow and should lead to a fault free installation.

1. From the 3 amp spur/socket use a 0.75 mm flex to the wiring centre.

2. Carry out the four part Electrical Safety Tests on the wiring centre.

3. Turn off the power to the wiring centre after the safety tests and check the wiring centre is isolated i.e. 0 Vac between live and neutral.

4. Wire the programmer L N E into the wiring centre, followed by any other components that require L N E connections such as the boiler, pump and room stat.

5. Connect the CH ON from the programmer to the wiring centre, joining it to the room stat COM.

6. Connect the room stat CALL to the wiring centre, joining it to the Brown/White/Brown+White of the motorised valve.

7. For **Honeywell two port systems**: Connect the HW ON from the programmer to the wiring centre, joining it to the cylinder stat Terminal1. For **all other systems**: Connect the HW ON from the programmer to the wiring centre, joining it to the cylinder stat COM.

8. For **two port systems** connect the cylinder stat CALL to the wiring centre, joining it to the Brown/White/Brown+White of the motorised valve. For **three port systems** connect the cylinder stat CALL to the wiring centre, joining it to the room stat CALL and Brown/White/Brown+White of the motorised valve.

9. Wire in the HW OFF if used.

10. Connect the orange wire(s) from the motorised valve to the boiler switched live/Pump.

11. For two port systems connect the grey wire(s) from the motorised valve to the wiring centre permanent live. For a three port system connect it to the wiring centre, joining it to the Programmer HW OFF.

12. Double check all neutrals and earth's are connected.

13. Power on the wiring centre, create a heating demand first and establish that the boiler has fired.

14. Turn off the room stat and ensure the boiler turns off.

15. Test that the programmer correctly turns on and off the heating demand.

16. Carry out the steps above to test the hot water.

Sequence of Operation

Always test in the heating mode because lots of boilers do NOT use a pump in hot water mode.

Some boilers run in low gas up to 3 minutes, the fan will also run slow in the HEATING start mode.

1. Check low gas pressure: If the ignition gas is too low the boiler will fail so you may need to increase the gas pressure. Seek guidance from manufacturer

2. A fan with 3 terminals (2 live + 1 neutral) has a pilot or runs in low gas in the 3 minutes start of heating mode. If not check for a blocked flue or venturi.

3. Firing sequences in heating mode:

   a: 100°C limit stat should have no resistance (buzzer sounds with multimeter

   on continuity mode or 0.0Ω on resistance setting).

   b: Pump – flow/pressure micro switch goes from NC to NO.

   c: Fan – fan switch goes to NC to NO.

   d; Temperature sensors show at least 3kΩ (3000 Ohms).

   c: Any stats will have no resistance (0.0Ω) except a frost stat and pump over run, which will show O.L.

4. Carry out the following basic ignition checks:

   a) Ensure a 3 amp fuse is fitted

   b) Check the polarity of the supply voltage is correct (see Electrical Safety Tests)

   c) Check the standing pressure is at least 20 mbar and the working pressure is over 18 mbar on the gas valve inlet

   d) Test the reset button

   e) If the boiler fails to fire up check gas valve has power (if no voltage is found work backwards to the PCB).

Lock Out or Failed Ignition

1. Low or no gas pressure inlet 20 mbar

2. Low or no water pressure 1 or 1.4 bar see page 76

03. Bent electrodes, replace

04. Over heated or high limit stat in O.L. should be NC

05. Fan switch not reset still in NC should be NO

06. Check electrode for spark

07. Check E to N on PCB must be 0 Vac makes strong spark

08. Check sensors and stats

09. Check pump voltage do NOT spin see page 38

10. Check fan voltage do NOT spin

11. Check the flue / fan Venturi is not blocked

12. Check condensing pipe for blockage or frozen

13. Check polarity see pages 16-17

14. Check pilot flame covers thermocouple

15. Replace thermocouple with original NOT universal

16. Check flame rectification don't, bend electrodes replace

17. Check Air Pressure Switch min 2 mbar do NOT blow

18. PCB damage, replace for the latest version

19. Check flue pressure

20. Check external controls boiler PCB voltage on N to SW.L

Noisy boiler

1. Pump is slow or 6 years old replace with latest version

2. Drain and flush with Fernox F3 then F2

3. Fit Fernox Electrolytic Scale Reducer

4. Fit Fernox TF1 Omega Filter

5. Gas pressure too high

6. No by-pass fitted

7. Inadequate by-pass fitted

8. Pump over run defective

9. Fan over run defective

10. Balance system

11. Faulty sensor / stat

12. Overheat stat faulty

Pressure Loss

1. Check Pressure Relive Valve

2. Check Auto Air Vent

3. Expansion vessel flat or too small

4. Rad valves leaking

5. Pipe connections

6. Wall stains

7. Boiler isolating valves

8. Inside the boiler for Leaks

9. Filling loop not capped off

10. Rads have been vented but not topped up

Popular Faults

1. **Cold rad across to top**: Turn off the boiler and pump, use tissue paper and SLOWLY undo the vent key until water shows do NOT over tighten, top up pressure to 1.0 bar.

2. **Hot rads in the summer or when CH is off:** Could be diverter stuck in heating mode or the boiler is in "pump over run" cooling mode after hot water draw off.

3. **Hot Water is not hot enough:** See 2 or sensor / stat fault or flow rate is too fast, see manual.

4. **Hot Water too hot or slow:** Dom. Exchanger is blocked (replace) cylinder stat is faulty, return water filter is blocked (replace) flush system check gas pressure is not too high call Gas Safe engineer.

5. **Some rads are hotter than others:** System rebalance call Gas Safe engineer, not a DIY job, check both rad valves are open, replace stuck Thermostatic Rad Valve.

6. **Losing pressure:** see overflow pipe form Pressure relief valve, see auto air vent, loose connections.

7. **Water pressure gauge is rising and falling:** expansion vessel needs recharging with air, or replacing and / or additional 12/15 ltr. to be fitted.

8. **The boiler ignites over night:** Frost stat activated 5-7°C

Popular Faults

9. **Rads are hot upstairs cold downstairs**: Replace pump and check expansion vessel.

10. **Rads hot downstairs cold upstairs:** Vent rads with the pump OFF, Check loft F&E tank has water, if no replace ball valve if yes there is an air-lock see cold feed pipe has a 3°deg fall.

11. **Boiler not responding to room stat:** Power OFF, Check in buzzer mode connection between room stat and SW.L.

12. **No display:** Check 3 amp fuse and 230Vac voltage on PCB Power OFF call factory for guidance.

13. **No Heating:** Diverter stuck in HW mode replace, room stat set too low, programmer in OFF mode

14. **No Hot Water:** Diverter stuck in CH mode or broken flow switch or sensor(s) Replace.

15. **Keeps blowing 3amp fuses:** Power OFF call factory for guidance.

16. **Intermittent heating:** low water pressure, electrodes, Pump, fan, air pressure switch, gas pressure, PCB,

FERNOX
MAKES WATER WORK

The noise you hear is directly related to the rapid condensation or implosion of stem in water. Localised boiling develops on the surface of the lime-scale deposit, which creates small steam bubbles. It is this movement of bubbles that creates the rattling noise as it leaves the deposit surface and travels round the central heating system.

Gas Valve
Only Gas Safe to test
Do NOT repair replace with newest version

1. Check voltage to M.I.

2. Gas pressure inlet 20 mbar working 18mbar or call manufacturer

Measure pins 1 to 5 and 1+2 and 4+5,
3 is earth seek manufacture guidance

Stuck or sized: Do not hit Replace

Letting by: Burner does not go OFF, power OFF then gas OFF, call manufacturer

If you smell gas, don't switch anything OFF or ON tell others open windows exit the building and call 0800 111 999

Mr Combi Fault Finding Checklist 'A'

Symptom - No Central Heating / No Domestic Hot Water

1. Carry out preliminary electrical safety checks

2. Switch main electrical supply off for 30 seconds and then switch back on. This will clear any lock out or central heating anti-cycling devices (some boilers have a lock out reset button)

3. Select central heating mode

4. Check external controls are calling and that there is power to the boiler

5. Check any internal frost stats if fitted.

6. Check the gas supply to the boiler

7. Check the water in the system. Se check list "D"

8. Check the overheat stat (should be manual re-set)

9. Check that the pump is running

10. Check the differential pressure valve, if fitted

11. Check the ignition system:

Fully Automatic	Pilot Type
Check the operation of the fan and the airflow sensor. Check the ignition	Check that the pilot is established. *Remember that the overheat thermostat could be a thermocouple interrupter type*

12. Check the gas valve

13. Check the boiler thermostat

14. Check the thermistor(s). See thermistor guide

15. Check the printed circuit board, fuses and connectors

Mr Combi Fault Finding Checklist 'B'

Symptom - No Central Heating / Domestic Hot Water ok

1. Carry out preliminary electrical safety checks

2. Switch main electrical supply off for 30 seconds and then switch back on. This will clear any lock out or central heating anti-cycling devices (some boilers have a lock out reset button)

3. Check the water in the system. **See check list "D"**

4. Select Central Heating mode

5. Check the external controls are calling and that there is power to the boiler

6. Check the pump (On twin pump models check the C/H pump)

7. Check the boiler stat or thermistor. **See thermistor guide**

8. Check the diversion medium is set the correct position and all switches have operated

9. Check the domestic hot water supply for leaks or dripping taps

10. Check that any domestic hot water pre-heat system is not operating and keeping the boiler in the hot water mode

11. Check the printed circuit board, fuses and connectors

Mr Combi Fault Finding Checklist 'C'

Symptom - No Domestic Hot water / Central Heating OK

1. Carry out preliminary electrical safe checks

2. Switch the main electrical supply off for 30 seconds and then switch back on. This will clear any lockout or heating anti-cycling devices. Some boilers may have a lock out reset button.

3. Check the mains water supply

4. Check the pump

5. Check the hot water over heat stat if fitted

6. Check the hot water stat or sensor

7. Check the hot water pre-heat if fitted

8. Check the printed circuit board, fuses and connectors

Mr Combi Fault Finding Checklist 'D'

Symptom - Sealed System
Under-pressurised or Over-pressurised

Systems are usually pressurised to approximately 1 bar. (15 lbs./ sq. inch)Accurate figures can be obtained from the manufacturer's instructions.

Low Water Pressure in System

- **Check** for water leaks.

- **Check** for micro water leaks. Question the customer as to when the system was last topped up.

- **Check** that the customer is not venting the radiators unnecessarily.

- **Check** for discharge at the safety valve inlet If the safety valve itself is in order then any discharge would indicate that the expansion vessel is under-pressurised of that the boiler is overheating. One other extreme possibility is that the diaphragm in the expansion verified by signs of water at the schraeder valve when it's released.

- **Check** the air pressure in the expansion vessel must is the <u>same</u> as the fill pressure i.e. 3 bed or smaller is 1 bar, larger systems 1.4 bar. all must NOT rise above 1.8 bar See chapter 'Expansion Vessel'. **Note:** when making this check the boiler must be drained to release the pressure on the water side of the diaphragm.

- **Check** that the pressure gauge is working. This will be achieved by adding or draining some water off, then checking the reading on the gauge.

High Water Pressure in System

- **Check** for filling loop left connected and passing water.

- **Check** the calorifier for break down, mains water passing into the primary water. This can be checked by firstly ensuring that the filling loop is disconnected. Then switch off the power supply to the boiler. Close the CH flow and return valves, then reduce the pressure in the boiler by draining the CH water off. Any rise in pressure, which will register on the gauge, indicates that the calorifier is leaking internally.

- **Check** that the pressure gauge is working. This will be achieved by adding or draining some water off, then checking the reading on the gauge

Printed Circuit Board

The job of the fuse

Inside the fuse is a special type of wire which is designed to melt at its rated current flow, thereby breaking the circuit and protecting the boiler from damage. However, it takes a small amount of time for this to happen, dependent on how much current is flowing and the rating of the fuse fitted.

Whilst it might only take a fraction of a second for the fuse to blow, you might be surprised how much damage can be done in that time. The PCB below was inside a boiler which was incorrectly fitted with a 13 amp fuse:

The electrical power consumption of a boiler is pretty low, so it only needs to be fitted with a 3 amp fuse.

If a higher fuse is fitted you should replace it immediately

PCB upgrades

Manufacturers regularly release new versions of boiler PCBs to add extra functionality and to improve circuit design. Newer boards are often much more reliable than older models so it's always a good idea to check if a replacement is available and suggest replacing it to the customer in order to make their boiler more reliable.

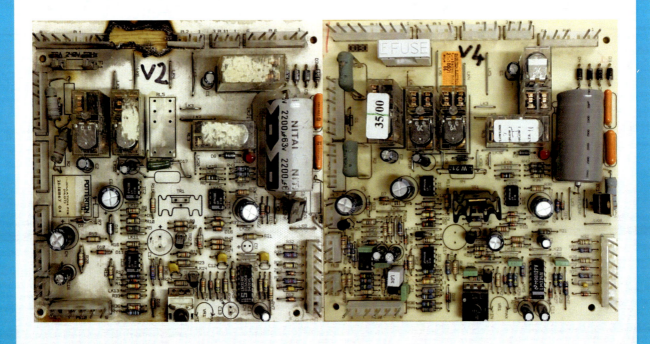

Look at the difference, more relays and components for the same boiler and just one year later. Do NOT fit recon's as a permanent solution.

The version number should be stamped on the board, often on the bottom:

When replacing a PCB, never fit reconditioned and always buy new from an approved stockist. ***Reconditioned PCB's can only be fitted as a temporary measure whilst waiting for the new replacement.***

PCB testing

We can test a PCB to check it's serviceable. To carry out these checks the PCB must be disconnected and removed from the boiler.

1. Check the condition on both sides looking for any signs of overheating, loose or missing components.

2. Short circuit test:

   a) Select the Resistance (Ω) setting on the multimeter.

   b) Place the black lead on the neutral terminal and the read lead on live.

   c) The resistance should be at least 100Ω. A lower resistance may indicate a fault, please contact the manufacturer for guidance. 0Ω indicates a short circuit.

3. Resistance to earth test:

   a) The multimeter should remain in the Resistance (Ω) setting.

   b) Place the black lead on the earth terminal and the red lead on the live terminal. Earth to Live must be O.L. Earth to Neutral must be O.L.

If any of these checks fail contact manufacture for guidance. If any of these checks fail, contact the manufacturer for guidance.

Pumps

NEVER remove the screw to spin the impeller to free when stuck, it's a **venting** screw.

Most boilers are fitted with a Grundfos 5 or 6 metre pump (designed for a 5-6 metre building), set to maximum speed. You do NOT need to change the whole unit just use a 'T' bar or 4mm Allen key to change the motor. In general, the higher the resistance (Ω) the slower it turns.

Testing the pump

There are two tests required for a pump using a multimeter:

1. Disconnect the pump from the PCB. Measure the resistance between Live (L) and Neutral (N). For a large 3/4 bed house it should be 150-230Ω, on a small heating system the resistance could be as much as 300-400Ω. If the resistance is 150Ω or less the pump should be replaced. If the resistance is too low it means the pump is running at a higher speed. This is less effective at moving the water and means the pump will overheat (this does not apply to 'A' rated pumps which are multi-speed).

2. Measure resistance between Live (L) and Earth (E) followed by Neutral (N) and Earth. The resistance should be very high (1MΩ) or open line (O.L.), if there is no resistance (0.0Ω) the pump will cause damage to the appliance and should be replaced immediately.

All pumps are rated by their power consumption in watts (W). A strong pump will be around 120W and a weak one 50W. When replacing a pump, ensure the correct power rating is fitted or the pump will not run at the correct speed. An 'A' rated pump should always be used on open systems. If you replace a pump, measure the resistance between the Live and Neutral pins and write the value onto the pump with a permanent marker. This will aid future fault finding.

On larger systems you should recommend to the customer that they change from a three port to a two port system.

  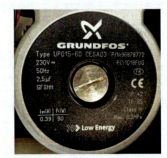

NOTE: 15.50/60 pumps have been discontinued, contact the manufacturer for guidance.

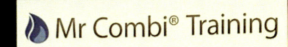

Balancing Radiators

Please Note, any system that does not have an automatic by-pass fitted within 1.5 metres from the boiler valves will be difficult to balance.

Method 1: With the boiler OFF.
Tools needed, small adjustable spanner, black permanent marker pen.

Turn OFF every lock-shield valve, mark the position, slowly turn and count and note how many turns to max.

The largest rads this is the setting 100% ON, medium size rads closer to the boiler 50% ON all the small Rads 25% ON.

Method 2: With the boiler ON.
Tools needed: multimeter,
Turn ON to MAX every control valve including TRVs, turn the room and boiler stat to MAX,
Wait 10 minuets to warm up, list all the rads, **warning HOT surface** touch the temperature probe top centre of each rad to show which rad(s) are over heating. Turn down those lock-shield valve(s) a little at a time to get a more even temp in deg.C

Method 3: With the boiler ON
Tools needed: Laser thermometer, recommended, as method 2.

Please note, the display is not the actual surface temperature.

Thermostatic Radiator Valve

TRVs increase comfort levels and save lots of money with a small outlay. Best position is on the FLOW diagonal to the return.
Upgrading to wireless Smart TRVs saves even more money.

Pumping Over & Static Head

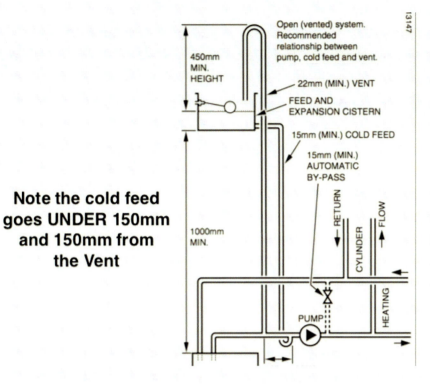

Open (vented) system.
Recommended
relationship between
pump, cold feed and vent.

l3147

450mm MIN. HEIGHT

22mm (MIN.) VENT

FEED AND EXPANSION CISTERN

15mm (MIN.) COLD FEED

15mm (MIN.) AUTOMATIC BY-PASS

RETURN

CYLINDER

FLOW

1000mm MIN.

HEATING

PUMP

Note the cold feed goes UNDER 150mm and 150mm from the Vent

22mm is a better cold feed

Extend over 2 meters
to stop 'pumping' over

Note the vent pipe has been cut above the water level and bent 75 deg. And extended 2 mtrs. A new 22mm pipe will be added to bring back to the F&E tank. Extra thick pipe lagging and support will required

We do not recommend soldering in the loft area push fit or compression.

By Pass

Location in System

System with stored hot water

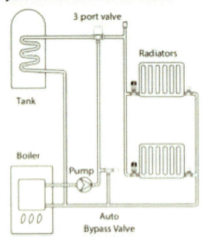

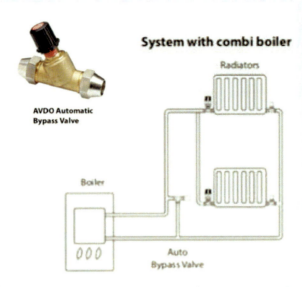

AVDO Automatic
Bypass Valve

System with combi boiler

Building Regulation Requirements

Government approved guidance covering the part of the Building Regulations related to the conversion of energy states that in all systems in which the boiler manufacturer's instructions demand a bypass, or specify a minimum flow rate through the boiler, then an automatic bypass valve be fitted and the manufactures instructions on minimum pipe length be followed. This type of valve modulates open when necessary to ensure that the appropriate minimum flow is maintained through the boiler, at all other tiles it is closed thus preventing unnecessary and wasteful circulation through the bypass and the boiler.

In addition to providing boiler minimum flow control , automatic bypass valves can be used to reduce persistent water velocity noise that may occur in some systems

Functions

1. Boiler manufactures will require a constant flow and pressure to be maintained.

2. In pump over run mode an auto by-pass will alleviate a pressure and noise problem.

Domestic Heat Exchangers

Plated heat exchangers

The heat exchanger has alternating plates passing cold water in one direction and the heated water from the boiler (82°C) in the other direction.

A 24kW boiler will have 12 plates. The flow rate is 8-10l/min.

A 28kW boiler will have 16-18 plates. The flow rate is 10-12l/min.

A normal bath filled to just under the overflow holds 180 litres. Therefore it should take 18 minutes to fill from a 24kW boiler and 12 minutes from a 28kW boiler. Always check the bath and sink tap with a proper Flow Cup.

If the flow rate is correct, gas burner/working pressure are also correct at least 18 mbar, then hot water should be okay. Heat exchangers rarely scale, 99% sludge up, therefore fit a filter on the heating return

Heat exchanger faults

Check for the following faults associated with the heat exchanger:

1. The flow rate is too low or the boiler fires up then goes off (hot/cold) water: The exchanger is 'sludged' up due to a dirty boiler/ system. Power flush, de-scale and replace exchanger. Fill with Fernox F2.

2. In water mode you get kettleing, banging etc: The exchanger is 'sludged' up again power flush as above and replace exchanger.

3. Back flow (the boiler fires up then goes off):

   a) Turn off the cold feed to the boiler.

   b) Open sink hot water tap, check there is no flow and then turn off.

   c) Check any blending tap/shower, move the lever to HW max.

   d) Any flowing water is a back flow fault. Replace the cartridge and/or fit non return valves.

De-scaling a heat exchanger is not recommended, always fit new. Flush the boiler and fill with Fernox F3.

If you replace a heat exchanger tell customer there is NO guarantee unless a return filter is fitted Fernox TF1. Inform them that muck in the rads will block the new exchanger.

Before fitting a new boiler and removing the old one you should power flush. Always fit a good inline anti-sludge unit like Fernox TF1 on the return and fill with an Fernox F2 inhibitor.

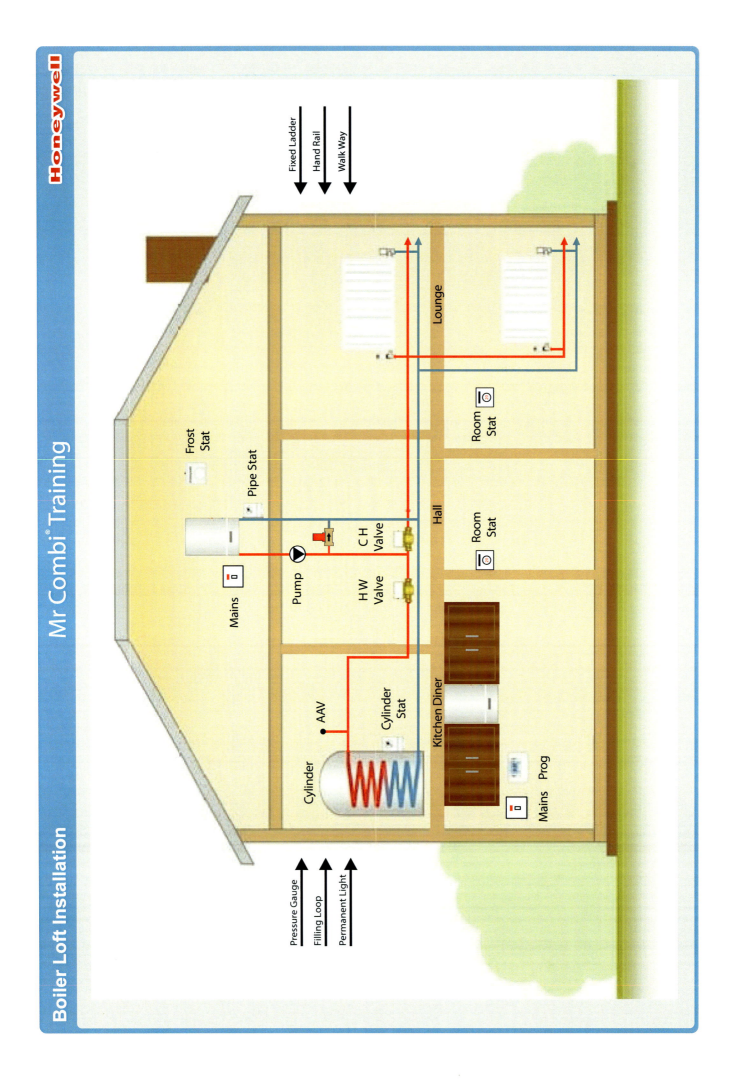

Loft & unheated area Installations

BS. 6798 : BS. 5422 : BS. 5970

BPEC CEN 1 Guidance regarding unheated installation of gas boiler is;-

1. Loft flooring must be provided to the boiler for service access.

2. Boiler has to be mounted on wall capable of withstanding weight.

3. Loft to be accessible with retractable ladder.

4. Must have fixed lighting inplace.

5. The roof space exit must be protected with a guardrail.

6. Gas, water, and electrical isolation points should be provided outside of the roof space so boiler can be isolated without gaining access to roofspace.

7. Condensate from the combi must be drained effectively and efficiently. A frozen condensate drainage pipe will prevent the boiler operating. Ideally, this pipe should been closed within the property to prevent freezing during very cold weather. Often in sufficient consideration is given to this potential problem and a condensate pipe is run down an exterior wall. This pipe should be insulated. If it should freeze, an engineer might refuse to climb an external ladder to remedy the problem due to insurance considerations.
It is therefore advisable to insulate all exposed pipe-work in a loft as a precaution against frost.

8. The boiler must have a factory fitted frost protection system.

9. The installer must fit a secondary frost protection system to protect the pipes and condensing pipe.

10. A pressure gauge and filling method must also be fitted outside the unheated area.

End user should be aware

11. Gas usage will rise and the direct debit will change accordingly

12. Hot water draw off will increase significantly and the direct debit will also change

Diverter Valves

The job of the diverter value is to direct (divert) the very hot water (82°C) from the main exchanger to either hot water exchanger or the radiators. It is usually made of brass it has a piston and spring mechanism which shunts back and forth when the hot tap is turned on. However it normally rests in heating mode so it can dump excess heat on the rads.

Some newer boilers will have a small motor (activated by a water flow switch) to push the piston, but most will use a rubber diaphragm that has a 5 year life. The complete diverter valve has a life expectancy of 7 years.

Faults in HW mode

Check for the following faults in hot water mode:

1. Water is warm and not HOT. Check the heating flow pipe, if it is heating up (creeping heat): The valve is 'letting –by'.

2. There is a hot water demand, the boiler fires up but the rads get hot and you get cold water: The valve is stuck in CH mode and should be replaced.

3. The hot tap is turned off but the boiler keeps working and is noisy: This means the diverter valve is not disengaged from the micro switch. Also check the HW flow switch is off.

4. Hot tap is turned on and nothing happens: Check the diaphragm, water flow switch etc.

Faults in CH mode

Check for the following faults in central heating mode:

1. Heating is required, the boiler fires up then goes off, pumps still runs: The diverter is stuck in hot water mode.

2. Boiler does not respond: The water flow switch is stuck or broken. Do NOT replace just the diaphragm the fault is the rear so change the whole diverter, charge £300 and repressurise the expansion vessel.

S PLAN FAULTS & TESTS
All testing MUST be made by a suitably qualified electrician

Many faults are due to programme set up errors. First clear the memory & reprogram as new. Make sure the Grey wire is on 1 & always Live 240V ac. All neutrals MUST go to 2, all earths MUST go to 3. With the WHITE 28mm or 1inch V4043H valve, the wire is not used & MUST be made safe.

No Heating or Hot Water:
1 Check the 3amp fuse is OK

2. Display is showing CH or HW (ON) if not programmer is faulty

3. 240v AC on the boiler and pump Live terminals, if no volts go to 4.

4. 240v AC on the Orange wire 10 if no volts go to 5.

5. 240v AC on 8 HW which goes to cylinder stat C if no volts go to 6.

6. 240v AC on 6 wire from the clock HW (ON) replace programmer.

Test Room Stat:
1. Turn stat down to the lowest temp.

2. Turn (OFF) the power

3. Remove and make the wire safe on T3

4. Turn ON power, programmer only to CH (ON)

5. Multimeter to AC black lead to T2 neutral red to T1 = 240 Vac

6. Put red lead to T3 = 0 Vac if more replace, should be dead.

7. Turn stat to maximum, must be the same as T1 if not replace stat.

No Hot Water: Test Cylinder stat:
1. Turn stat down to the lowest temp.

2. Programmer only to HW (ON)

3. Multimeter to AC black lead to EARTH red to T1 = 240Vac

4. Put red lead to C = 0 Vac if more replace, should be dead.

5. Turn stat slowly to maximum, must be the same as T1 if not replace stat.

6. Must be fitted 1/3 up from the floor set to 60°C

Test Zone Valve: (Boiler & Pump dead):
1. Programmer to HW or CH (ON)

2. Multimeter to AC black lead to EARTH

3. Red to 5 CH or 8 HW = 240 Vac if yes replace motor

4. If the motor works but the boiler is still dead check 10 Orange = 240Vac and 1 Grey = 240Vac if yes replace zone valve.

5. If the boiler and pump continue to run when the clock / stat says (OFF) replace zone valve.

Wired Sundial S Plan & S Plan Plus

The table opposite gives guidance on a quick electrical check for installed wired **Sundial S Plan** and wired **S Plan Plus** to help in commissioning and to pin-point the source of any electrical problems.

Remember the **Golden Rule** when you have a problem. First of all **check your wiring.** Only start suspecting faulty components after you are satisfied all wiring is correct.

The following notes will help to identify faulty components.

Cylinder Stat
First of all, make sure you have wired to the correct terminals.

Terminal C (common) is the **Left Hand** terminal.
Terminal 1 is the **Middle** terminal.
Terminal 2 is the **Right Hand** terminal.

Suspect the cylinder thermostat is faulty only if Terminal C **not** live when calling for Hot Water.

Room Stat
1) Remove wire from Terminal 3.
2) Live to Terminal 1.
3) Turn stat to call, if no live on 3 then faulty.

Suspect the room stat is faulty only if Terminal 3 is not live when calling for heat. (Make sure Terminal 1 is live). While checking, disconnect wiring from Terminal 3 to prevent false readings due to backfeed.

Zone Valves
Suspect a motorised valve is faulty only:

1. If the motor fails to rotate with live applied to the **Brown** wire and neutral to the **Blue** wire. (Motor can be viewed with valve cover removed).

    Note that the motor stops automatically when the valve is fully open and stays in this condition as long as live is applied to the **Brown** wire.

    The valve automatically closes under spring return when live is removed from the **Brown** wire.

2. The **Orange** wire only becomes live after the valve has fully opened (Make sure the **Grey** wire is live).

3. If the boiler and pump continues to run when the cylinder stat and room stat is satisfied and the clock is in OFF position.

Programmer
Suspect the programmer only:
(a) After you have made sure that any links required are in place.
(b) After you have made sure that the Programmer has power – to the correct Terminal.
(c) After you have made sure that the Programmer timing is set up correctly (see individual Programmer User Guide as appropriate).
(d) If live does not appear at Heating ON Terminal when Heating is selected on continuous or timed.
(e) If live does not appear at Hot Water ON Terminal when Hot Water only is selected on continuous or timed.

Wired Sundial S Plan & S Plan Plus

Programmer Switch Position	Heating only selected	Hot Water only selected	Hot Water and Heating selected
Programmer	Live on 'CH ON' Terminal.	Live on 'HW ON' Terminal.	Live on both 'HW ON' & 'CH ON' Terminals.
T6360B Room Thermostat	Set to call for Heat. Live on Terminals 1 & 3.	No live on any terminal.	Set to call for Heat. Live on Terminals 1 & 3.
L641A Cylinder Thermostat	No live on any terminal.	Set to call for Hot Water. Live on Terminals C and 1.	Set to call for Hot Water. Live on Terminals C and 1.
V4043H Heating Zone Valve	Live on **Brown**, **Grey** and **Orange** wires.	Live on **Grey** and **Orange** wires.	Live on **Brown**, **Grey** and **Orange** wires.
V4043H Hot Water Zone Valve	Live on **Grey** and **Orange** wires.	Live on **Brown**, **Grey** and **Orange** wires.	Live on **Brown**, **Grey** and **Orange** wires.
Boiler and Pump	Boiler and pump fired via live feed from **Orange** wire.	Boiler and pump fired via live feed from **Orange** wire.	Boiler and pump fired via live feed from **Orange** wire.

NOTES

1. Check **must** only be made by a suitably qualified electrician.
2. **Grey** wire on both Heating and Hot Water zone valves **must** be connected to permanent live.
3. **Blue** wire on both Heating and Hot Water zone valves **must** be connected to neutral.
4. Terminal 2 on room thermostat **must** be connected to neutral.
5. Ensure that any links required in programmer are in place.
6. Earth connection (**Green/Yellow**) **must** be made on valve.
7. With 28mm or 1 inch V4043H valves the **White** wire is not used and **must** be made electrically safe.

SEE NOTES OPPOSITE IF YOU HAVE A PROBLEM

Honeywell Y Plan
Backfeed

FAULT: 50Vac or more on the white wire to neutral, the boiler may fire up with no demand, fit this snubber between boiler Live or (SW.L) and Neutral on the terminal strip

Roxburgh EMC XEB Series Mains Protector, Flange Mount Mounting

RS Stock No.: **672-7035**   Mfr. Part No.: **XEB1201**   Brand: Roxburgh EMC

✓ **3135 In stock for FREE next working day delivery**

Price Each (In a Pack of 5)
£4.972   **£5.966**
(exc. VAT)   (inc. VAT)

Units	Per unit	Per Pack*
5 - 20	**£4.972**	**£24.86**
25 - 120	£3.55	£17.75
125 - 245	£2.79	£13.95

The maximum heating load is 13KW recommended 10KW of rads. A 3 bed house may require over 15KW thus a 2 zone (S Plan) should be installed

Honeywell   3 Port valve

Heat capacity		Size of pipes (mm)	Size of valve (mm)	
Kw	Btu/Hr		3 Port	2 Port
13	45,000	22	22	22
23	80,000	28	28	22
36	125,000	35	N/A	28
51	175,000	42	N/A	2 x 22
73	250,000	54	N/A	2 x 28
(All figures quoted at atmospheric pressure)				

Y PLAN FAULTS & TESTS
All testing MUST be made by a suitably qualified electrician

Many faults are due to programme set up errors. First clear the memory and reprogram as new. All neutrals MUST go to 2 an then all earths MUST go to 3. Now check the fuse 3 amp.

Cylinder Stat - Test

Turn down to the lowest temperature. Power OFF to disconnect & make safe wires T1 & T2 (this stops back feed) Power ON Programmer set only to HW (on) multimeter black lead to EARTH Red lead to 6 test 240 Vac if dead replace P/gram.

1. Check Com = 240V ac left hand side

2. Slowly turn up stat T1 = 240 Vac also 8 = 240 Vac if not replace stat

3. T2 = 240V ac only when stat is satisfied normally 0V ac

Test Room Stat:

1. Turn stat down to the lowest temp.

2. Turn (OFF) the power

3. Remove and make the wire safe on T3

4. Turn ON power, programmer only to CH (ON)

5. Multimeter to AC black lead to T2 neutral red to T1 = 240 Vac

6. Put red lead to T3 = 0 Vac if more replace, should be dead.

7. Turn stat to maximum, must be the same as T1 if not replace stat.

Mid-position Valve - Test (heating mode):

Suspect the V4073A valve is faulty only does not operate as specified in the following checks (These should be done in order1. 2. 3. 4. 5 and 6)

1. Turn (OFF) the power

2. Disconnect Grey 7 and White 5 wires from junction box reconnect to permanent Live 1 Turn ON power, motor will go to CH (ON) port A check Orange 6 =240Vac no? Replace valve

Mid-position Valve - Test (hot water mode):

1. Switch off mains the valve should now automatically return to open HW port B and close heating port A

2. Switch off mains, isolate Grey and White wires and make safe, remove cylinder stat wire 6 and reconnect to permanent Live 1, switch ON turn stat to 60°C the pump and boiler should start.

Mid-position Valve - Test (hot water & heating)

1. Switch off mains replace the wire to 6 put Grey 7 and White 5 wires from junction box reconnect to permanent Live 1 Power back (ON) both stats to max the valve goes to MID position.

2. Switch off mains replace Grey 7 and White 5 wires valve is OK, wiring fault

Programmer - Test

1. Check all links are in place

    b. check wiring is correct

    c. check programmed OK

    d. live does not appear on CH continuous

    e. when live does not appear on HW continuous

    f. when live does not appear on HW (OFF)

Wired Sundial Fault Finding

The table opposite gives guidance on a quick electrical check for installed wired **Sundial Y Plans** to help in commissioning and to pin-point the source of any electrical problems.

Remember the **Golden Rule** when you have a problem. First of all **check your wiring.** Only start suspecting faulty components after you are satisfied all wiring is correct.

The following notes will help to identify faulty components.

Cylinder Stat
First of all, make sure you have wired to the correct terminals.

Terminal C (common) is the **Left Hand** terminal.
Terminal 1 is the **Middle** terminal.
Terminal 2 is the **Right Hand** terminal.

Suspect the cylinder thermostat is faulty only if Terminal 1 does **not** become live when calling for Hot Water, or Terminal 2 does **not** become live when satisfied. (Make sure that Terminal C is live in both cases). While checking, disconnect Terminals 1 and 2 to prevent false readings due to backfeed.

Room Stat
1) Remove wire from Terminal 3.
2) Live to Terminal 1.
3) Turn stat to call, if no live on 3 then faulty.

Suspect the room stat is faulty only if Terminal 3 is not live when calling for heat. (Make sure Terminal 1 is live). While checking, disconnect wiring from Terminal 3 to prevent false readings due to backfeed.

Mid-Position Valve
Suspect the V4073A valve is faulty only if the valve does not operate as specified in the following checks (these should be done in order 1, 2, 3, 4, 5 and 6).

Valve open for Heating only
1. **Switch off mains supply.** Disconnect **Grey** and **White** wires from appropriate junction box terminals. Reconnect both **Grey** and **White** wires to permanent live terminal in junction box.
2. Switch on mains supply. Valve motor should now move to fully open heating Port A. The motor should stop automatically when Port A is open, and stay in this position as long as power is applied to **White** and **Grey** wires. When Port A is fully open, the **Orange** wire becomes live, to start pump and boiler.

Double check by feeling that Port A outlet is getting progressively warmer.

Valve open for DHW only
3. Switch off mains supply. The valve should now automatically return to open DHW Port B and close Heating Port A.
4. Isolate **Grey** and **White** wires and make safe. Remove cylinder stat wire from Terminal 6 in junction box and connect to permanent live. Switch on fused spur, cylinder thermostat must be set to call for heat, pump and boiler should start.

Valve open for both DHW and Heating
5. Switch off mains supply. Replace cylinder stat wire to Terminal 6. Isolate and make safe **Grey** wire and connect **White** wire to permanent live. Switch on mains supply, motor should now move to mid-position and stop automatically. Cylinder thermostat must be set to call for heat. Both ports A & B are now open for Hot Water and Heating. Boiler and pump should start.

Double check by feeling that pipe outlets from ports A & B become progressively warmer.

6. Switch off mains supply, reconnect **White** and **Grey** wires to junction box terminals.

If this check completes satisfactorily, the problem is not the valve, but elsewhere in the circuit.

Programmer
Suspect the programmer only:
(a) After you have made sure that any links required are in place,
(b) After you have made sure that the Programmer has power – to the correct terminal,
(c) After you have made sure that the Programmer timing is set up correctly (see individual Programmer User Guide as appropriate),
(d) If live does not appear at Heating ON Terminal when Heating only is selected on continuous or timed,
(e) If live does not appear at Hot Water ON Terminal when Hot Water only is selected on continuous or timed,
(f) If live does not appear on Hot Water OFF terminal with Hot Water OFF on programmer.

Wired Sundial Fault Finding

Programmer Switch Position	Heating only selected	Hot Water only selected	Hot Water and Heating selected
Programmer	Live on both 'CH ON' & 'HW OFF' Terminals.	Live on 'HW ON' Terminal.	Live on both 'CH ON' & 'HW ON' Terminals.
T6360B Room Thermostat	Set to call for Heat. Live on Terminals 1 & 3.	No live on any terminal. (See note 2 for Terminal 3).	Set to call for Heat. Live on Terminals 1 & 3.
L641A Cylinder Thermostat	Nominal 90 volts. Live on Terminals 1 & 2 (Note Terminal 1 only becomes 240 volt live after V4073A valve opens and Boiler fires). (See notes below).	Set to call for Hot Water. Live on Terminals C & 1. (See note 2 for Terminal 2)	Set to call for Hot Water. Live on Terminals C & 1. (See note 2 for Terminal 2)
V4073A 3 Port Mid-Position Valve	Live on **Grey, White** and **Orange** wires. Valve opens to Port A for Central Heating (CH).	Live on **Orange** wire only (See note 2 for **Grey** and **White** wires) Valve not energised. Port B open for Domestic Hot Water (DHW).	Live on **White** wire and **Orange** wire. (See note 2 for **Grey** wire). Valve in mid position for CH and DHW.
Boiler and Pump	Boiler and pump fired via live feed from **Orange** wire.	Boiler and pump fired via live feed from Terminal 1 on cylinder stat.	Boiler and pump fired via live feed from Terminal 1 on cylinder stat and **Orange** wire.
NOTES	1. Check **must** only be made by a suitably qualified electrician. 2. Low A.C. voltage may appear on specified wire or terminals due to back feed from V4073A valve. If in doubt, disconnect **Grey** or **White** wire as appropriate, or check with meter for full 240V. 3. **Blue** wire on valve **must** be connected to neutral. 4. Terminal 2 on room thermostat **must** be connected to neutral. 5. Ensure that any links required in programmer are in place. 6. Earth connection (**Green/Yellow**) **must** be made on valve. 7. Earth connection not needed on room stat or cylinder stat.		

SEE NOTES OPPOSITE IF YOU HAVE A PROBLEM

Y Plan Valve Operation

Voltage Readings at Valve

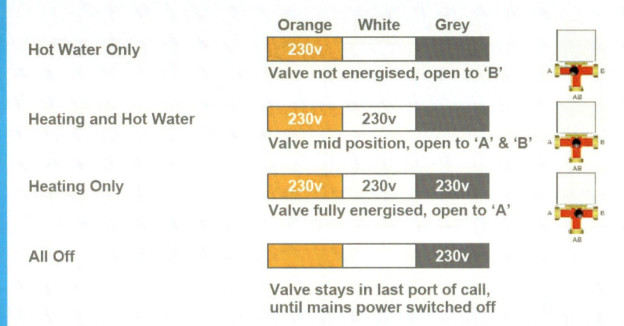

	Orange	White	Grey	
Hot Water Only	230v			
Valve not energised, open to 'B'				
Heating and Hot Water	230v	230v		
Valve mid position, open to 'A' & 'B'				
Heating Only	230v	230v	230v	
Valve fully energised, open to 'A'				
All Off			230v	

Valve stays in last port of call,
until mains power switched off

Fault finding: Here we have test voltages that can be measured to check the valve operation at each stage. The V4073A has 3 ports - to ensure the valve is fitted round the right way a simple way of remembering is: Port A is for AIR (Heating), Port B is for BATH (Hot Water).

When you buy the valve in its box and before it is connected to anything there are 3 ports but only 1 ball - Where is the ball?

Answer: Against Port A (spring return). The valve is open to the Hot Water port.

When checking a valve that is installed, to establish where the ball is, the process should always be:

• Switch off the fused spur isolator,
• Turn the Programmer off for both channels,
• Turn both thermostats down. Now you know that the ball is relaxed, closed to Port A, open to B, for Hot Water.
• Now turn the fused spur isolator on
• Turn the Programmer on for both channels
• Turn the Cylinder Thermostat only up to calling. This will fire the boiler, whether the motor in the valve is working or not.

Checking Individual Components Part 1

Programmer

First check that any links required are in place (for volt free programmer switches). Check that there is a live and neutral mains supply. Disconnect the output terminals (CH on, CH off, HW on, HW off) to ensure that false readings are not being obtained from back-fed signals, especially in three port valve systems. The method of checking the programmer is then to ensure that it is powered up, i.e. mains available, and them check in turn the terminals CH on, CH off, HW on, HW off to see if 240V is present when the programmer is switched to the relevant mode of operation.

Thermostats

In three port systems especially, the output terminals of the thermostats are subject to back feed from the motorised valves. Thus check the operation correctly the output terminals must be disconnected. Once this is done, with a mains voltage on the supply to the thermostat (easiest done by leaving existing programmer on constant and checking live feed at the thermostat), move the thermostat until it clicks to indicate operation of the internal switch and test for the correct voltage at the correct terminal:

Turn room stat right down, The room stat satisfied (satis) should be live (not all room thermostats have this terminal).

Turn room stat right up, The room stat call should be live.

Turn cylinder stat right down, Thermostat satisfied (satis) should be live (note: cylinder thermostat may not be able to turn off if cylinder stone cold on some models).

Turn cylinder stat right up, Thermostat call should be live.

Frost Thermostats

Sometimes systems can be left running continuously because the frost thermostat has been mistaken for a room thermostat and turned up to around 20°C. Check the setting. Disconnect any outputs and then test for operation as above.

Checking Individual Components Part 2

Motorised Valves (spring return only)

2 Port Motorised Valves: Check motor runs when power applied to brown wire with blue wire connected to neutral. Check motor returns to closed position under spring pressure when brown wire disconnected i.e. it is not jammed open. Check that end switch is functional: ensure permanent live to grey wire available. Orange wire should become live when motor fully open and microswitch closes. Check that microswitch similarly opens when motor winds back. To distinguish between faults on valve body (valve stuck or sticking) and faults with electrical actuator for types with removable head recheck operation with valve head removed. If actuator now works satisfactorily now renew valve body since seized.

3 Port Mid Position Valve: (4 or 5 wire types only) These have three checks, one for each position: If valve does not function correctly, is at fault and should be replaced:

1 Test Heating Only: Isolate from mains. Disconnect grey and brown (white in some valves) wires from valve and connect both to permanent live. Reconnect to mains. Motor should run to fully open position (heating only) and orange wire become live. To confirm, leave boiler (and pump) connected to orange, when boiler fires only heating port should become hot.

2 Test Hot Water Only: Isolate from mains. Valve should motor back under spring return to block heating port completely. To confirm position of valve: disconnect grey and brown (white) wires and make safe. Connect boiler (and pump) live to permanent live. Reconnect to mains. Boiler should fire and only hot water port should become hot.

3 Test Mid Position: Both heating and hot water. Isolate from mains supply. Disconnect grey wire and make safe. Connect brown (white) wire to permanent live. Reconnect to mains. Valve should motor to mid position. Confirm be connecting boiler (and pump) to permanent live and checking both ports become hot. To distinguish between faults on valve body (valve stuck or sticking) and faults with electrical actuator for types with removable head recheck operation with valve head removed. If actuator now works satisfactorily now renew valve body since seized..

T4360 Frost Thermostat and
L641B Pipe Thermostat

To reduce the risk of frozen pipe work during severe cold weather, Frost Protection can be installed to protect either the whole central heating system or the boiler and localised pipe work. These controls are designed to override the Programmer and Room Thermostat controls whether wired, wireless or wireless enabled.

If a Frost Thermostat only is to be installed to protect the whole central heating system , it must be sited where a rise and fall in heated air temperature can be detected , i.e. in a room with a radiator, and set to 12-16°C. This function is built in to programmable thermostats and Sundial RF². If the Frost Thermostat is installed outside the heated area, i.e. in a boiler room, garage or attic space, it is strongly recommended that a Pipe Thermostat be used as well to ensure that overheating of the property does not occur. The Frost Thermostat should be set to 5°C. The Pipe Thermostat will sense a rise in water temperature in the pipe work and then switch the system off. It should be sited on the boiler return, set at 25°C and wired as below.

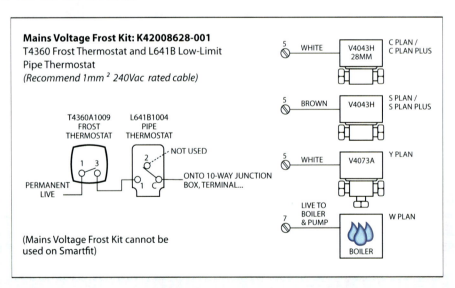

When a Frost Thermostat is installed on a central heating system, the fused spur should only be switched off for servicing and maintenance. If the heating system is to be switched off for any other reason, eg. holiday, then switching must only be carried out at the Programmer or Time switch, otherwise the Frost Protection is disabled.

Frost and Pipe Thermostat Positions

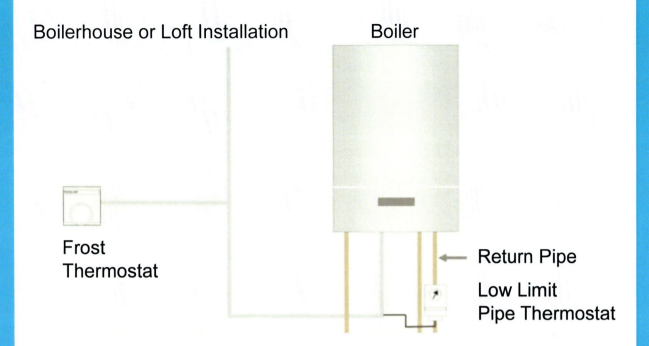

Boilerhouse or Loft Installation   Boiler

Frost
Thermostat

← Return Pipe

Low Limit
Pipe Thermostat

When the boiler is installed in an unheated area, e.g. the garage or loft space, then the boiler and pipes are at risk from freezing conditions, so they need frost protection.

Traditionally a Frost Thermostat was fitted immediately above the boiler or between the flow and return pipes. Here it would pick up enough heat so that would be enough heat to switch off after a short time.

Unfortunately a high efficiency boiler does not let any or much heat out and the pipes are lagged so the boiler could remain on for hours or even weeks before the temperature rises above 5°C. Inside the house the heating will be on continuously.

To avoid this problem, fit a two stage frost protection. All this requires is the addition of a pipe stat.

When the Frost Thermostat detects freezing conditions (i.e. below 5°C) it switches on the Pipe Stat set to 25°C fitted to the return pipe. This will detect a rise of temperature when the pipework has been warmed through and switch off the boiler.

The Frost Thermostat should always be fitted where the risk of freezing occurs

FAQ Sheet

Positioning a room thermostat correctly

The performance of all room thermostats is affected by the air flow across them which they measure. This air flow is dependent on the location of the room thermostat. If a room thermostat is poorly located, the air flow will not be representative of the rest of the room, and the temperature control will be adversely affected.

Because every heating system must have a room thermostat, the decision of where to position it is very important. There are places where a room thermostat should never be found. There are places which need careful consideration before installing a room thermostat. Finally, the places that are left will be good for locating the room thermostat.

Where you should put it
The diagram below gives good guidance on where a thermostat should be fitted.

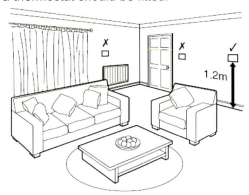

Locating a room thermostat

Generally, it is very difficult to suggest the perfect position, as every heating system is different. Locate the room thermostat in the heated area (zone) requiring control where it has a free flow of air around it on wall at a height of about 1.2m. But, do make sure that the thermostat is not suffering from any of the adverse factors in the following list.

Where you should not put it
It is, however, very easy to eliminate all of the bad places to site the room thermostat.

Do not fit a room thermostat:

- In a room with another major heat source, e.g. an open fire, gas fire or cooker
- In an unheated room
- In a room fitted with radiator thermostats
- In direct sunlight
- Behind furniture or curtains
- In a warm draught
- In a cold draught
- Directly opposite a radiator, or other heat source
- Directly above a radiator, or other heat source (Don't forget that electrical appliances emit considerable amounts of heat. E.g. Television, DVD Player, Hi-fi etc
- In a corner of two walls
- In a corner at the junction of the wall and ceiling

Where you need to think about it
Some positions for a room thermostat may be perfectly acceptable, but exceptional problems may need to be considered.
- On an external wall. The room thermostat may be on a cold wall, therefore overheating the living space. (Cure: Turn room thermostat down.)
- On a garage wall. Sometimes an electrician may surface mount the cable in the garage and then drill through the wall to access the back of the thermostat. This can allow a very cold draught directly into the back of the thermostat, reducing the sensed temperature, therefore causing serious overheating of the living space. (Cure: Seal the hole with filler.)

heatingcontrols.honeywellhome.com

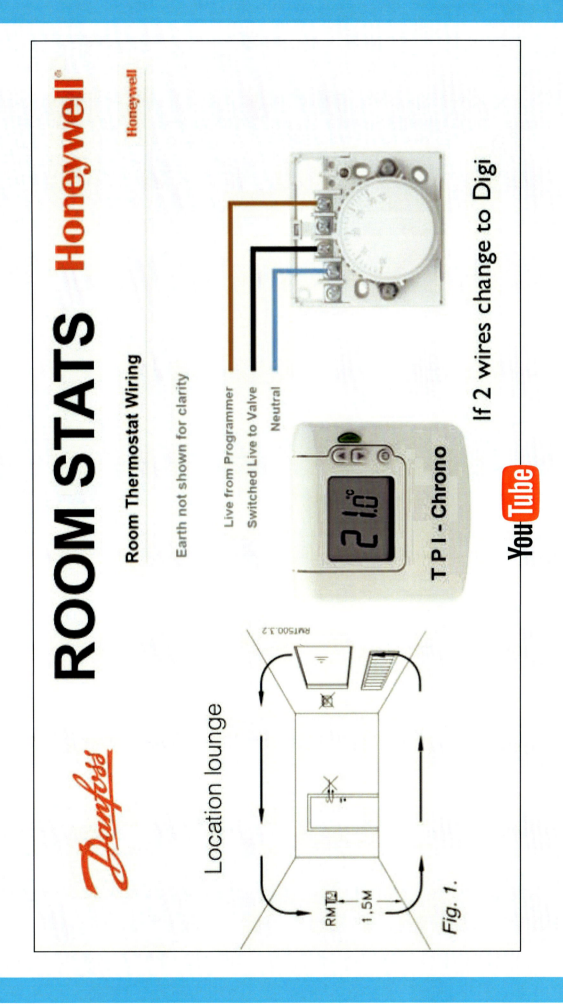

ROOM STATS

Honeywell

Danfoss

Honeywell

Room Thermostat Wiring

Earth not shown for clarity

Live from Programmer

Switched Live to Valve

Neutral

T P I - Chrono

If 2 wires change to Digi

Location lounge

RMT500.3.2

RMT

1,5M

Fig. 1.

YouTube

Cylinder Thermostat Wiring

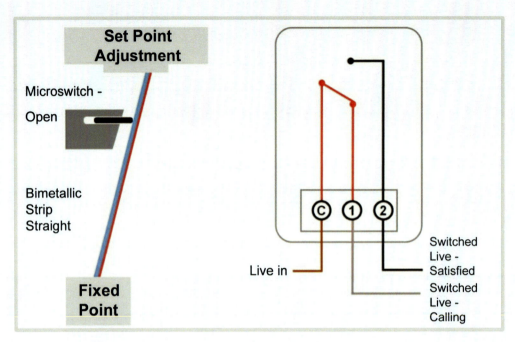

On most cylinders a rapid acting thermal response is not required and the sensing element can be less sensitive. Thus we use a bimetallic strip. These are two dissimilar metals with different coefficients of expansion bonded together.

When cool the strip lies straight, when it is heated, it bends. As it bends it hits a micro switch to switch OFF the heat to the cylinder.

When the stored hot water is up to temperature (60°C) the heat from the water is transferred to the bimetal strip. Because both ends of the bimetal strip are fixed the strip bends hitting the micro switch causing the contacts to change over, i.e. terminal C now makes to terminal 2 (satisfied).

When the cylinder thermostat is satisfied at the top limit, it will not switch on again until there has been (approximately) a 10°C fall.

With the bimetallic strip straight (calling) the switch is made C-1. When the bimetallic strip is bent (satisfied) the switch is made C-2.

If we only require a Calling signal, its safer electrically to connect Live in on 1, Live Out on C. By doing this terminal 2 never becomes live.

If we need both a Calling and a Satisfied signal we will show the cylinder thermostat with live on C then a calling Live is on 1, satisfied Live out is on 2.

Remember this for simplicity; In on C, call on 1, Satisfied on 2.

Cylinder Thermostat
Where and Why?

- A wide differential ensures that the boiler is not fired for a small demand

- Positioning affects volume of stored hot water

- Too high, not enough Hot Water

- Too low, cylinder thermostat does not switch off

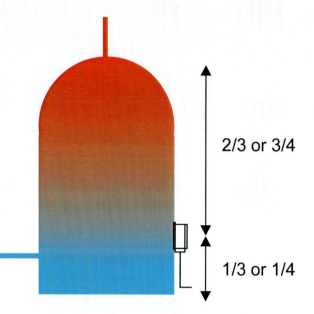

2/3 or 3/4

1/3 or 1/4

This slide shows the correct position for a Cylinder Thermostat. Cold water enters the bottom of the cylinder and when heated being less dense rises to the top (from there it goes out to the taps when they are opened).

As the hot water is drawn off then the cold water flows in at the bottom. If the Cylinder Thermostat is positioned too low this means it senses cooler water in the cylinder too quickly and results in excessive boiling water cycling.

If the Cylinder Thermostat is positioned too high then is satisfied when only a little water has been heated, so there is not sufficient hot water storage.

A Rapid Recovery Cylinder require rapid acting aquastats (these normally have a probe into the cylinder) with a differential set to 4°C. These cylinders usually hold less volume of water and have a greater surface area of heat exchanger (coil).

They need the fast reacting thermostat to avoid overheating. Many cylinders of this type have many metres of small diameter copper coil. This transfers heat to the small volume of water very quickly. The height on the thermostat is dictated by the position of the "well" provided by the cylinder manufacture.

Hot water problems

Check voltage to switch

1. Often low DC volts, to test, turn OFF power In ohms (rest mode between Com and NO must be O.L turn the tap on, reading must change to 00.0 ohms any other reading it's faulty

2. Check the cold water flow rate is more than 3 ltrs. more than the combi flow rate boiler output

3. Check water flow is more than 4 ltrs. min. diverter is not in CH mode.

4. Some plate exchangers have a filter, wear gloves, if blocked Flush the whole system replace filter and plate exchanger fill with Fernox F1 or we recommend F2.

5. On a tank / cylinder system check the cartridge in the tap.

6. Cylinder has furred up, replace.

7. Filters are blocked.

8. Switch stuck ON in the off mode.

Water Pressure & Flow Switches

Switches can be 2 or 3 pin and all switches (air, gas & water) have the same connection names:

C – Common - This pin is supplied with voltage from the PCB

NC – Normally Closed - The NC terminal has a connection to the Common whenever the switch has not been made

NO – Normally Open - The NO terminal is connected to the Common when the switch is made, this removes the connection to the NC terminal

Most switches will have a small diagram showing which pins are which:

NOTE: some manufacturers will reverse the NC and NO to power frost stat or a pump overrun stat.

Always test switches on boilers in heating mode because the pump always works and does not upset the customer by leaving a tap running.

Testing without power applied

You can test a switch directly with your multimeter on the resistance setting.

Place the black lead on the COM and the red lead to check at the NC, with the switch open there should be a short circuit (0.0Ω). Between COM and NO there should be an open circuit (O.L. or a straight line). When the switch is closed these readings should be reversed. If the multimeter shows any numbers during these tests the switch may be faulty.

Testing with power applied

When testing a switch with the appliance switched on, you shouldn't use the resistance setting on the multimeter, you should use the AC voltage setting instead.

Place the black lead on the boiler casing (earth) and use the red lead to measure the voltage at the COM, NC and NO terminals in turn. There should be between 50V and mains voltage at the COM, the same voltage (see note) should be found at the NC terminal when the switch has not been made and there should be 0V at the NO terminal.

The voltages at the NC and NO terminals should reverse when the switch is made.

NOTE: If the voltage found on the NC or NO terminal is less than at the COM, this is caused by damage to the switch terminals (due to arcing). The switch should be replaced.

Fans

Testing the fan

There are two electrical tests for fans using a multimeter:

1. Measure the resistance between the two pins, the resistance should be between 30 and 80Ω. If the resistance is less than 20Ω or more than 100Ω the fan should be replaced.

2. Measure the resistance between each pin and earth (the casing of the boiler should be earthed as it will have been checked in the electrical safety tests). The resistance should be very high (1MΩ) or open line (O.L.), if there is no resistance (0.0Ω) the fan will cause damage to the appliance and should be replaced immediately.

Some fans have a two-speed motor and therefore have 3 pins. The slow speed is for a pilot version or in heating mode when the boiler fires on low gas for 3 minutes. When carrying out the resistance test place the black lead on the Neutral and measure the other two pins in turn. The pin that shows the higher resistance is for the slower fan speed.

All boilers with a pilot will have a slow running fan, this is controlled either directly from the PCB or from an in-line gold resistor. Use caution as the resistor can be extremely hot and is easy to damage.

Single Speed Fan

Two Speed Fan

Physical checks

Carry out a physical check of the fan. Look up the chamber for any black fluff or leaves.Remove the flue / elbow for a thorough clean. Spin the blades, they must rotate freely, slow down gently and be free from noise, otherwise the fan should be replaced.

Use extreme caution when handling the fan, it is very easy to unbalance you could be replacing it for free. If the fan is over 6 years old it should be replaced, along with the fan switch.

Air Pressure Switches
NEVER EVER blow into an Air Pressure Switch!

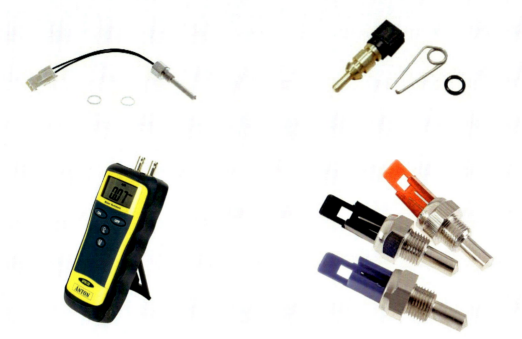

All testing should be done with a Cat.3 multimeter and on the PCB

Function: To check the flue and main heat exchanger are clear and safe for ignition.

Many modern boilers may have a sensor

On arrival (cold)

Measure the resistance between both terminals to the M.I. 00.0 Ω or O.L. is a fail. The PCB will send a signal to check resistance before ignition then within 30 seconds re-test to see the difference is within the permitted values, if not the boiler will shut down and a Fault Code may be displayed as "blocked flue". With the power OFF measure the APS **Com** (black lead) to **NC** (red lead) reading must be 0.00 Ω and **Com** to **NO** must be O.L. anything else is a fail, the APS has not reset, replace with the latest version.

Test with the fan on:

Before switching ON disconnect the positive tube from the Venturi found on fans or flue elbows and put into a digital manometer, test on heating ON mode (the boiler can't fire up but may lock-out within 30 seconds) to test the positive pressure is over 2 mbar. Look up into the flue bend for the presence of leaves, black "fluff" blockage. Ventures can melt or block. Manufacturers will change parts if found to be unreliable.

Temperature Sensors & Thermostats

Stats

A thermostat is a preset temperature switch (on/off) at a given temperature (HW 60°, CH 80°, limit 100°C) or it will switch on a pump to cool the boiler down below 60°(pump overrun) if no demand is on.

Sensors

A sensor changes resistance as the temperature rises or falls in the boiler. This goes to the modureg on the gas valve to adjust the gas flow to modulate from low to high flame or OFF. Most sensors range between 13kΩ cold to 2kΩ hot.

At room temperature (20°C) it should read about 10-13KΩ and as the temp rises it usually goes down to 2KΩ. However some may start low then go up to 13KΩ or more. The important thing to check is that the resistance of the sensor changes as the temperature changes. If this does not happen sensor is faulty, it could be scaled up or O.L. (open circuit). The heating sensor is most commonly at fault; if two are fitted always replace both.

Testing sensors

A problem with testing sensors is how to measure it. On some boilers if you disconnect the leads to measure the resistance across the pins the boiler will either shut down or go to high flame and then over heat. Testing needs to be quick or a better testing place needs to be found such as on the PCB. It's useful to have some spare new sensors for diagnostics.

A better solution:

1. Fit a manometer to the burner test nipple.
2. With the boiler cold, disconnect the wires from sensor(s), measure the resistance and make a note of the value.
3. Refit the wires, switch on the boiler to maximum CH temp. for 2 minutes.
4. Quickly turn down the stat and note change of gas pressure (some boilers stay on low gas up to 3 min).
5. Switch off the boiler, disconnect the sensor leads and measure the sensor resistance again, there should be a big drop from the first measurement.

On our DVD this subject is covered in greater detail because this IS very complicated and difficult to understand

Thermistor Testing Guide

Various types of thermistor sensor are used on combi boilers and different testing methods could be used to check them out. These checks fall into three patterns and are listed below as A, B and C. To test out a suspect thermistor look on the following boiler charts to see which check is required (A, B or C) and then carry out the appropriate check on the thermistor.

Test A

- Isolate the boiler from the electrical supply
- Pull the leads off the thermistor
- Reconnect to electrical supply
- Switch on the boiler, in the faulty mode CH or DHW
- The burner should fire
- If the burner fires, the thermistor is faulty, replace the thermistor
- If the burner does not fire then the fault lies elsewhere

Test B

- Isolate the boiler from the electrical mains
- Connect the thermistor leads together
- Reconnect to electrical supply
- Switch on the boiler in the faulty mode CH or DHW
- The burner should fire
- If the burner fires the thermistor is faulty, replace the thermistor. If the burner does not fire then the fault lies elsewhere

---

Test C

- Isolate the boiler from the electrical mains
- Pull the leads off the thermistor
- Read the resistance across the thermistor terminals and check the results against the boiler charts. If the thermistor is outside these limits then replace it

Or

- If the boiler has thermistors for CH and DHW cross connect the suspect thermistor leads to the other thermistor, switch the boiler on in the faulty mode, if the burner fires then the thermistor is faulty, replace the thermistor. If the burner does not fire then the fault lies elsewhere

Thermistor Testing Guide

Vaillant NTC

Sensor

Stat

Sensor & Thermistor Resistances

Temperature (°C) Tolerance ± 10%	Resistance (kΩ)
20	14.772
25	11.981
30	9.786
35	8.047
40	6.653
45	5.523
50	4.608
55	3.856
60	3.243
65	2.744
70	2.332
75	1.99
80	1.704
85	1.464
90	1.262
95	1.093
100	0.95

Vaillant NTC sensors

Negative Temperature Coefficient (NTC) sensors on Vaillant boilers usually have one pin so when testing place the red lead on the pin and the black lead on the boiler casing, it should read approximately 3.4KΩ.

Pilot Light & Thermocouples

Pilot light fault

If the pilot light goes out after 2 minutes, check the low and high burner pressures. The boiler could be over heating usually in hot water mode, check the heat exchanger(s) pump speed, sludge, flue / chamber ingress, (broken seal).

Thermocouples

The thermocouple is a safety device that stops gas escaping when the pilot light goes out.

Honeywell Q309

Universal
Not specified for sale

Commons faults

Two common faults associated with the thermocouple are where a pilot light won't light and where the pilot light goes out when the gas valve button is released. To resolve this issue carry out the following steps:

1. Check the gas supply is over 18mBar (20-22 is better).
2. Check the flame completely covers the tip, if not clean / replace the pilot jet.
3. Test the Hi Limit stat 100°C (continuous buzzer on multimeter or 000.0! replace if over 6 years old).
4. Replace the thermocouple with the manufactures original (use a Uni whilst on order). If you see a Uni fitted, order a new original.
5. Replace the Gas Valve.

Wire Wound Resistor (250-900Ω)

The WWR is activated by the boiler flow / pressure 50mm diaphragm, a pin moves to activate a micro switch about 160vAC which keeps the fan rotating at slow speed to provide air for the pilot.

These are fitted in boilers with a Fan Flue. DANGER, they are very HOT and brittle, most have been discounted due to age and not produced any more

DO NOT TOUCH as the wires are silver soldered and cannot be repaired, if the condition is poor order a new one, don't touch it. If the wire disconnects you MUST shut down the boiler otherwise in OFF mode the boiler will over heat and could explode..

Expansion Vessels

How expansion vessels work

As water heats up, it expands. In a sealed system this causes problems because the expansion will produce an increase in pressure. In a heating system the water can expand by as much as 4%, so an expansion vessel is fitted to absorb the extra volume. The vessel is in two halves; one half fills with water from the system and the other half is filled with air at the same pressure as the system. A rubber diaphragm separates the two halves. When the central heating is running, the water heats up, it expands and the air in the vessel is compressed to allow room for the extra water volume. As the system cools, the water volume shrinks and the air is able to expand back to its original volume. If the vessel is the right size and is filled to the right pressure, the system pressure should never rise above 2.5 bar. Over-sizing the vessel helps to reduce the rise in pressure, which is good for the system. It also helps with future proofing the house, for example if larger or extra radiators are fitted.

Unfortunately, installers often don't check the EV pressure when they install a boiler. From the factory they are normally set to 0.5 bar or are flat.

Before installing a boiler, set the expansion vessel pressure in accordance with the instructions on the next pages

Expansion vessel faults

Faults caused by the expansion vessel are fairly common and should be easy to spot. If the expansion vessel is flat or too small, there is not enough room for the water to expand. As the systems heats up, the water pressure rises beyond 3 bar and is released from the Pressure Relief Valve (PRV).

As the system cools down, the water contracts and air is brought into the system via the Auto Air Vent (AAV) or the radiator valve glands. When the system next heats up this air will then escape back out of the AAV as the pressure rises.

Over time some of this air also collects in the radiators and then the radiators act like expansion vessels. The system will operate like this until the owner bleeds the radiators.

If you suspect that the EV is a cause of a problem, you must drain down the boiler to test the vessel pressure, otherwise they will be simply measuring the pressure of the system instead.

Common symptoms

Check for the following symptoms:

1. In heating mode does the pressure gauge rise over 2.5 bar and water drips from the PRV?

2. Check the AAV, does it show signs of leaking? Green or black stains?

3. Does the vessel have to be repeatedly refilled with new air, or does the pressure drop slowly, and customer finds no sign of a leak (could be a leak under the floorboards)?

If any of these symptoms exist, the expansion vessel pressure should be checked. Don't forget, the vessel pressure can only be checked once the boiler has been drained down.

Vessel pressure

The expansion vessel should be set to the exact SAME pressure as the water in the system, this must be determined first. It is calculated as 1.0 bar, plus the static height of the house multiplied by 0.1.

The static height is the vertical distance from mid-way up the expansion vessel to the top of the system. If two vessels are fitted this will be the lower of the two.

Single storey house: In a bungalow the expansion vessel in the boiler is at a higher point than the tops of the radiators so the system pressure will be 1.0 bar.

Two storey house: In a typical house with rooms of 2.4m, the distance from the expansion vessel in the boiler to the tops of the radiators on the first floor will be approximately 2m.

$$1.0 + (2 \times 0.1) = 1.2 \; bar$$

Three storey house: In a typical house with rooms of 2.4m, the distance from the expansion vessel in the boiler to the tops of the radiators on the second floor will be approximately 4.4m.

$$1.0 + (4.4 \times 0.1) = 1.44 \; bar$$

Always consult the manufacturer's instructions to find the operating range of the appliance. This will normally be 1-1.5 bar.

How to re-pressurise the vessel

This should be carried out if changing a pump, diverter valve or heat exchanger:

1. Turn off gas / electric and do not touch the isolating valves, the boiler should air-lock. Drain down the boiler via the return drain-off nipple or the PRV. It should be turned slowly so the valve is jammed open to empty the boiler.

2. On the vessel, check the Schroeder valve with a pencil gauge, it should be flat. If any air remains, let it out from the valve.

3. Using a foot pump, SLOWLY pump air up to the same as the normal system pressure. As the vessel is filled with air, the water will be pushed out through the boiler, thus flushing it. If water comes out of the Schroeder valve don't panic, it's condensation and it should be there.

4. Continue pumping, all the water will go, until the gauge reaches the desired pressure. Finally, check the valve for leakage.

5. Turn on the filling loop valve to 1 bar to clean the boiler from the inside. It may need 2-3 attempts. Once complete, if there are any brown or green stains on the AAV or if the vent cap has jammed, replace it. If necessary, replace the PRV also.

6. Close the drain-off, fill the boiler with liquid X200 / Silencer, then add water to the desired pressure.

7. Switch on heating mode, followed by HW mode, to remove air from system.

Fitting an external vessel

The only way to correctly size an expansion vessel is by working out the total water content for the system and then referring to a sizing chart such as below. To do this the initial system pressure must be determined first then the correct column can be selected.

Water content (litres)	System Pressure		
	1.0 bar	up to 1.2 bar	up to 1.5 bar
20	2.2	2.5	3.1
40	4.4	5.0	6.2
60	6.5	7.5	9.4
80	8.7	10.0	12.5
100	10.9	12.5	15.6

If the vessel in the boiler is not adequate for the system an external vessel will need to be fitted to make up the difference. It must be fitted on the heating RETURN pipe, as close to the boiler as possible, with 22mm copper pipe. Tests have shown that the vessel can satisfactorily be fitted up to 3 metres away via 22mm pipe.

NOTE: Both vessels must be pressurised to the same as the working pressure.

Many central heating systems have under-sized expansion vessels so this is a great money earner and should be charged at about £200.

Expansion vessel diagram

A Appliance expansion vessel

B Extra expansion vessel
   (central heating return)

C Drain cock

P Pressure relief discharge

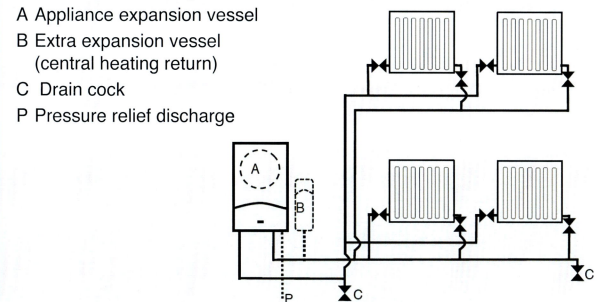

As above use a pressure adjusting valve, The minimum pressure is 0.75 bar

A 3 bed house or smaller, a sealed combi system the cold fill pressure should be 1.0 bar for a large property fill to 1.4 bar. No system should rise above 1.8 bar. The expansion vessel may need re-perssurising or replacing.

To stop 'over pressure' a pressure adjusting valve should be used set to 1bar or 1.4 for larger systems over 3 bedrooms.

This pressure adjusting valve is very useful for installers, saves time in filling & venting, by turning the red fill valve ON the engineer does not need to worry about over pressurising the system. This method is particularly suitable in rented accommodation, offices, shops where this could also be a problem.

Checking the CO2 & Ratio

This chart is only available from https://www.mrcombi.com/shop/

 Mr Combi©Training

CO/CO2 Ratio for Nautural Gas v2
Copyright Anton amendment by George Staszak Mr Combi ® Training 2018

%	%	CO ppm														
Oxygen	CO2	10	50	100	150	200	250	300	350	400	450	500	550	600	650	700	750
3.0	10.2	0.0001	0.0005	0.0010	0.0015	0.0020	0.0025	0.0029	0.0034	0.0039	0.0044	0.0049	0.0054	0.0059	0.0064	0.0069	0.0074
4.0	9.6	0.0001	0.0005	0.0010	0.0016	0.0021	0.0026	0.0031	0.0036	0.0042	0.0047	0.0052	0.0057	0.0063	0.0068	0.0073	0.0078
4.5	9.3	0.0001	0.0005	0.0011	0.0016	0.0022	0.0027	0.0032	0.0038	0.0043	0.0048	0.0054	0.0059	0.0065	0.0070	0.0075	0.0081
5.0	9.1	0.0001	0.0005	0.0011	0.0016	0.0022	0.0027	0.0033	0.0038	0.0044	0.0049	0.0055	0.0060	0.0066	0.0071	0.0077	0.0082
5.5	8.8	0.0001	0.0006	0.0011	0.0017	0.0023	0.0028	0.0034	0.0040	0.0045	0.0051	0.0057	0.0063	0.0068	0.0074	0.0080	0.0085
6.0	8.5	0.0001	0.0006	0.0012	0.0018	0.0024	0.0029	0.0035	0.0041	0.0047	0.0053	0.0059	0.0065	0.0071	0.0076	0.0082	0.0088
6.5	8.2	0.0001	0.0006	0.0012	0.0018	0.0024	0.0030	0.0037	0.0043	0.0049	0.0055	0.0061	0.0067	0.0073	0.0079	0.0085	0.0091
7.0	7.9	0.0001	0.0006	0.0013	0.0019	0.0025	0.0032	0.0038	0.0044	0.0051	0.0057	0.0063	0.0070	0.0076	0.0082	0.0089	0.0095
7.5	7.6	0.0001	0.0007	0.0013	0.0020	0.0026	0.0033	0.0039	0.0046	0.0053	0.0059	0.0066	0.0072	0.0079	0.0086	0.0092	0.0099
8.0	7.3	0.0001	0.0007	0.0014	0.0021	0.0027	0.0034	0.0041	0.0048	0.0055	0.0062	0.0068	0.0075	0.0082	0.0089	0.0096	0.0103
9.0	6.8	0.0001	0.0007	0.0015	0.0022	0.0029	0.0037	0.0044	0.0051	0.0055	0.0066	0.0074	0.0081	0.0088	0.0096	0.0103	0.0110
10.0	6.2	0.0002	0.0008	0.0016	0.0024	0.0032	0.0040	0.0048	0.0056	0.0059	0.0073	0.0081	0.0089	0.0097	0.0105	0.0113	0.0121
11.0	5.6	0.0002	0.0009	0.0018	0.0027	0.0036	0.0045	0.0054	0.0063	0.0065	0.0080	0.0089	0.0098	0.0107	0.0116	0.0125	0.0134
12.0	5.1	0.0002	0.0010	0.0020	0.0029	0.0039	0.0049	0.0059	0.0069	0.0071	0.0088	0.0098	0.0108	0.0118	0.0127	0.0137	0.0147
13.0	4.5	0.0002	0.0011	0.0022	0.0033	0.0044	0.0056	0.0067	0.0078	0.0078	0.0100	0.0111	0.0122	0.0133	0.0144	0.0156	0.0167
14.0	3.9	0.0003	0.0013	0.0026	0.0038	0.0051	0.0064	0.0077	0.0090	0.0089	0.0115	0.0128	0.0141	0.0154	0.0167	0.0179	0.0192
15.0	3.4	0.0003	0.0015	0.0029	0.0044	0.0059	0.0074	0.0088	0.0103	0.0118	0.0132	0.0147	0.0162	0.0176	0.0191	0.0206	0.0221

WORCESTER 28Jr HE

We have to check the emission is to M.I. which is 9.8% 1st reading is 10.5% much too high. See the CO is 184ppm very close to the max 200ppm.

XS air is only 12.9 far too low (means the flame colour is dark blue and too big). The Ratio is OK at 0.0017, if no action was taken this boiler will soot up and have high running costs and a shot life. In just 3 minutes I have saved this boiler, see the O2 is 3.7; CO is 84; XS air 21.7 and CO2 is 9.8 target reached happy customer

CO₂ settings

Note. CO₂ should be measured after 10 minutes

Gas type	CO₂ setting maximum	CO₂ setting minimum
Natural gas	9.8% ±0.2	9.2% ±0.2
LPG	11.0% ±0.2	10.5% ±0.2

▶ Check CO is less than 200ppm.
▶ Measure the inlet pressure; it should be no less than 18.5mb for natural gas and 37mb for LPG.

Servicing can save money

We all know it can be challenging to convince homeowners of the benefits of proactive servicing: but it may soon become an annual requirement if the government takes a step towards exploring new fuels to heat our homes. Here, Worcester Bosch's Martyn Bridges outlines the steps that can help improve the health of a heating system.

In a bid to meet the country's 2050 carbon reduction objectives, the possibility of using alternative fuels – or at least a different quality of gas – is becoming more likely. As many of these solutions involve changing the chemical properties of the gas that we use, inevitably this will have an impact on the performance of our appliances, making the need for regular checking and servicing of a heating system imperative.

In the meantime, regular servicing remains optional rather than required, and it's up to gas engineers to advise homeowners on its benefits if they want to improve efficiency and ultimately keep their running costs down.

1 Check the water
Most homeowners regard the content of the pipework in their heating system as out of sight, out of mind; and don't pay much attention to keeping it in good condition. However, research has proven that heating systems heavily contaminated with magnetite, sludge and various other corrosion impurities can be up to 6 per cent less efficient – which can make a big impact on the efficiency of a boiler.

For homeowners with little regard for the inside of their heating system, gas engineers can show them exactly what is inside the pipes by using a turbidity tube. Not only is this a useful comparative device to monitor that the discharge (dump) water has successfully cleared during the power-flushing process, but engineers are using the tool more often to demonstrate the level of clarity of water in their customers' heating systems.

Because an older heating system is likely to have been connected to four or five different boilers in its lifetime, the gas engineer's first priority should always be to ensure the water in the system is clean and contains corrosion inhibitor. Ideally, this should be an annual check to make sure the system is not corroding and has enough inhibitor to stay protected.

If a customer is looking to invest in further protection for their heating system, adding a system filter will complement the cleansing process; ensuring the long-term protection of an appliance by collecting any contaminants that may not have been removed at the time or that may build up afterwards.

2 Check the appliance
With the pipework clean, the next step is to look at the boiler itself. Gas engineers should make sure that its combustion efficiency reads as per the manufacturer's instructions.

This should ideally be carried out once a year: any excess air or insufficient air going into the boiler will affect the efficiency of the burner. Measure the CO/CO_2 ratio of the flue gases and adjust the appliance accordingly. Many manufacturers ask for an air pressure test to be taken, generally on a test point on the fan and on the exhaust outlet. The measured pressure indicates the cleanliness of the heat exchanger and whether it needs cleaning.

3 Check the controls
Once the engineer has checked that the boiler and the water in it are clean and healthy, it then falls to accessories such as heating controls to make any additional efficiency gains. Any control system should obviously time the on/off periods of the system, with a room thermostat to turn off the boiler and pump when the desired room temperature has been achieved.

Smart controls can offer a large efficiency uplift even without any human interaction, by enabling the boiler to adapt its behaviour according to how a homeowner uses their heating

system. This makes smart controls well suited to people who are keen to save energy.

The use of weather compensation, where the control can access live weather data to influence the boiler output, and load compensation, which adjusts the boiler's output in line with the actual room temperature, will have a real impact on bills.

4 Check the parts
In addition to the general servicing steps that should be carried out on every heating system, there will be some specific requirements that vary. Manufacturers use a range of different components in their appliances, all of which have different parameters – a bit like the servicing requirements of two different car manufacturers.

With that in mind, gas engineers should be on the lookout for any 'anniversary part changes', as the proactive replacement of ageing components will keep the boiler ticking along nicely. This could involve changing anything from a seal to an electrode every year or two, and can be determined by referring to the manufacturer's guidelines.

Although annual boiler servicing is currently not a legal requirement, the benefits of maintaining an efficient and effective heating system are plain to see. Gas engineers, industry bodies and manufacturers must continue to take every opportunity to challenge homeowners' attitudes and reap the rewards of greater heating efficiency. ■
• www.worcester-bosch.co.uk

> "Gas engineers should be on the lookout for any 'anniversary part changes' as the proactive replacement of ageing components will keep the boiler ticking along."

Why not come on a course?

We now offer two courses where you can spend a great day with George Staszak who will help you learn fault finding techniques and understand the cause of common problems with boilers and control systems.

Mr Combi Training is a family run business which aims to provide a friendly and stress free environment where you can learn and develop your skills at your own pace. George will teach you how to test in a sequence using a digital multimeter and manometer. YOU do all the work learning in a great atmosphere! This is a 'hands on' day, no sleeping or boring lectures, only guides. Whatever your level, from no experience to advanced, you will be surprised by how much you can learn in just one day.

The Fault Finding & Multimeter Course is designed for installers who want to improve fault finding servicing and repairs for their customers, or maybe change from installing and move into a servicing career. It also caters for new students coming into the heating industry, who have little or no knowledge of Combi boilers.

Danfoss & Honeywell have given us special permission to provide you with technical support not normally available. We have spent over two years with both companies' amazing support to find a way to bring you two courses on one day. We've also employed a graphics expert to help us to change the standard black and white schematic drawings into beautiful colour, this makes them so easy to follow just like the London underground map.

As well as our fantastic courses, we also offer two great DVDs and apps so you can learn at home or on the go, at your own pace. Watch them over and over to maximise your learning and improve your skills.

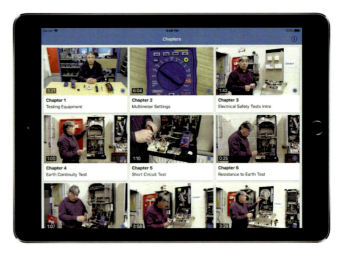

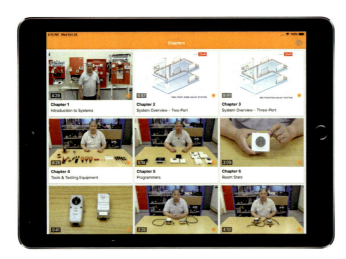

Search for Mr Combi Training on the App Store or the Google Play Store to find out more!

The aim of this book is to make wiring easier and more rewarding. It has been designed to help you carry out installations and fault finding on control circuits by giving you all the information you need in a clear and logical manner for gas installers and electricians with little or no experience of basic or smart controls.

We have worked closely with Drayton, NEST Danfoss and Honeywell to take their existing wiring diagrams and schematics and turn them into full colour illustrations which are much easier to understand.

The most common control circuits in use in the UK are the three port and two port circuits.

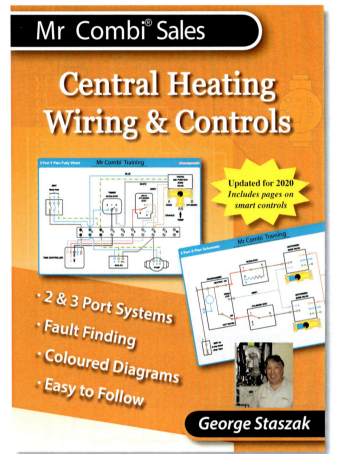
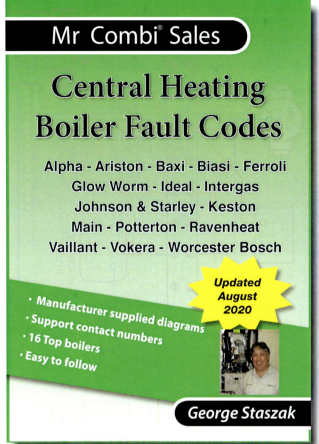

This book therefore contains a greater number of diagrams for those circuits.

• **Fully Wired** - Shows every wire in the control circuit
• **Live Wire** - The earth and neutral wire have been removed for clarity
• **Valve** - Shows the internal connections to the motorised valve
• **Schematic** - A simplified version of the circuit
• **Live Wire Stats** - Only the thermostat circuit is shown

We've also added plenty of other useful diagrams which we know are necessary when carrying out work on controls such as frost protection, by-passes, wall plates, fault finding flow charts and more.

Wiring centres hints and tips, Part L and P are explained, using a multimeter for Safe Isolation, **sequence of wiring,** why a by-pass is so important

There are pages on testing parts from Honeywell setting up a wireless thermostat to a receiver and lots more, our apps and hands on course details.

Fault Finding & Multimeters Course

If you install, service or repair Combis and want to learn easy fault finding, then this course is for you.

George will explain the sequence of operation, how to spot a faulty expansion vessel and refill them, [C - NC - NO] found on fan switches what do they mean? All will be revealed, pumps fans and sensors, learn the correct method to test them and lots more!

1. Why does the pressure rise to 3 bar and stay there? It should be 1.2 or 1.5 cold.
2. The pilot won't stay on but the Thermocouple is OK.
3. Hot water is OK but no heating, what to look for.
4. Fan runs, all seems OK but it just won't work?
5. Boiler fires up then goes out within 10 seconds - back flow?
6. Boiler blows its fuse! I will show you how to do the 4 part **Electrical Safety Tests** and find the fault safely, without using electricity.
7. We also cover temp sensors / stats and boiler service techniques.

You will be able to work on the boilers at your own pace so you will really learn a lot in one day. Start a new career there is a very good living to be made we are in demand! This is not a day off, lots of 'hands on'.

Spend a day with someone who has been repairing Combis for over 40 years, and still does. Learn how to use a multimeter and understand why 'Ohms' are so important.

We have over a dozen broken / faulty Vokera, Potterton, Worcester, Vaillant boilers for you to practice on at your own pace but always under supervision, use your own tools and spend a friendly informative day with us.

Testimonial

"Hi George, just wanted to thank you for a great day. It's a good feeling using my multimeter with confidence. You can pride yourself in your approach to training. Even with 10+ in the class you still had plenty of time for individual attention. Would recommend to any engineer/installer." Barry

DVDs & Apps

As well as our fantastic courses, we also offer two great DVDs and apps so you can learn at home or on the go, at your own pace. Watch them over and over to maximise your learning and improve your skills.

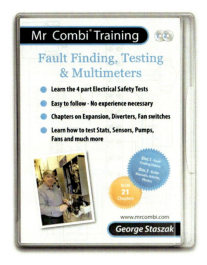 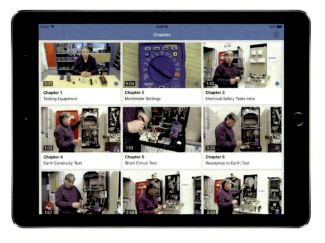

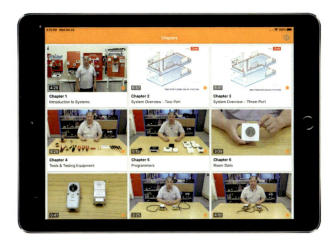 

Search for Mr Combi Training on the App Store or the Google Play Store to find out more!

Android Apps

Fault Finding, Testing
& Multimeters

Wiring & Controls

Wiring & Controls
- Diagrams

Gas Rate Calculator
& Guide

Heatloss Calculator
& Guide

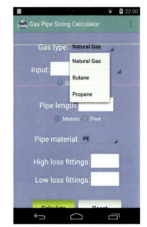

Gas Pipe Sizing
Calculator

Gas Ventilation Calculator

Expansion Vessel Calculator

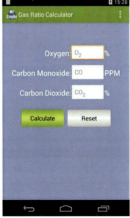

Gas Ratio Calculator